W9-CGY-150

The Mystery of the Mind

Books by Wilder Penfield

Scientific Studies:

Cytology and Cellular Pathology of the Nervous System, Editor, 1932.

The Cerebral Cortex of Man (with T. B. Rasmussen), 1950.

Epileptic Seizure Patterns (with K. Kristiansen), 1950.

Epilepsy and the Functional Anatomy of the Human Brain (with H. Jasper), 1954.

The Excitable Cortex in Conscious Man, 1958.

Speech and Brain Mechanisms (with L. Roberts), 1959.

Historical Novels:

No Other Gods, a story of Abraham, 1954.

The Torch, a story of Hippocrates, 1960.

Biography:

The Difficult Art of Giving, The Epic of Alan Gregg, 1967.

Essays:

The Second Career, 1963.

Science, the Arts and the Spirit, 1970.

The Mystery of the Mind

A Critical Study of Consciousness and the Human Brain

by Wilder Penfield, O.M., LITT.B., M.D., F.R.S.

Foreword by Charles W. Hendel
Introduction by William Feindel
Reflections by Sir Charles Symonds

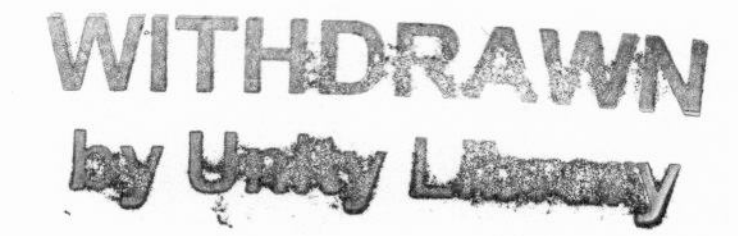

PRINCETON UNIVERSITY PRESS
PRINCETON, NEW JERSEY

10-85

Published by Princeton University Press, Princeton, New Jersey
In the United Kingdom: Princeton University Press, Guildford, Surrey

Library of Congress Cataloging in Publication Data will be found on the last printed page of this book

This book has been composed in Linotype Times Roman

Printed in the United States of America
by Princeton University Press, Princeton, New Jersey

First Princeton Paperback printing, 1978

Book store gift

Dedicated to

Sir Charles Sherrington, O.M., F.R.S.
physiologist and pioneer-explorer
of the nervous system

Contents

Preface

I began this writing when I was called upon to prepare an address for the annual general meeting of the American Philosophical Society held in Philadelphia, April 21, 1973, under the chairmanship of Francis Schmitt. It was subsequently elaborated a month later for the Hans Berger Centennial Symposium on Brain-Mind Relationships at the Montreal Neurological Institute.

When I reviewed the functional mechanisms of the human brain that could now be clearly recognized, I realized that the time had come to apply this knowledge in a modern reconsideration of the nature of the mind. I could not myself refuse to accept this exciting challenge, and so I turned away from other writing to the preparation of this monograph.

A Pilgrimage

Since I became a neurosurgeon, I have always been concerned primarily, of course, with the problem of one patient after another, and patients prompted many an inquiry. The patient continued to be in the foreground of my concern, but in the background there was an urge to exploration. It had been awakened in me first, I suppose, by the Professor of Biology at Princeton, E. G. Conklin. A little later, while listening to the lectures of Sir Charles Sherrington, as a medical student in Oxford, I realized that there was a thrilling undiscovered country to be explored in the mechanisms of the mammalian nervous system. Through it, one might approach the mystery of the mind, if only one could deal with the human brain as Sherrington had analyzed the reflexes of the animal brain.

And so, after my initial university training, I returned to Oxford for graduate work in laboratory neurophysiology and then went on blindly toward the human brain through neurology and neurosurgery. Thus, there was always a restless wondering within me about the working of the brain and its relation to mind. In time then I tried being an explorer, too, and stumbled, perhaps by accident, on much good gold.

If, in making this final report of my experience, I fail to describe adequately the excellent contributions of other explorers in the field, I can only apologize. We are joined in a common cause, and the primary duty of each of us is to give an account of his own intriguing expeditions into this undiscovered country and to pass on his own record accurately and hopefully to others who follow on these explorations that have now become a fateful pilgrimage.

Presentation to the Public

Philosophers of a certain school would, no doubt, silence me before I began to discuss the mind and the brain, if they could. They declare that since the mind cannot, by its very nature, have a position in space, there is only one phenomenon to be considered, namely, the brain. Such a declaration, which is contrary to the thinking of most men today as it was in ages past, is an unproven hypothesis. Like all such hypotheses, one should undertake to prove, or to disprove it, without initial prejudice.

A scientist examines the evidence before he presumes to draw a conclusion. This is particularly important when facing such an ancient (and such an important) question! Is the mind merely a function of the brain? Or is it a separate but closely related element? A physiologist can examine the brain. He has, as yet, no direct approach to the mind. That there is the closest relationship, however, is self-evident. Must the physiologist with facts at hand forever stand apart from the philosopher?

In the presentation of this book to a public of widely varied backgrounds, I have asked three distinguished friends from three different disciplines to join me in the publication: William Feindel, a neurosurgeon; Charles Hendel, a philosopher; and Sir Charles Symonds, a neurologist. Since I turned to Hendel as to a philosopher for assistance in the composition of the manuscript, let me explain our relationship: He and I graduated from Princeton University together in the Class of 1913. Indeed, I myself struggled with the philosophy of Kant sitting beside Hendel in the same weekly "preceptorial." In 1934 he and I, following very different routes, had come to live in Montreal. The Montreal Neurological Institute of McGill University opened its doors that year and as its director and a neurosurgeon, I looked forward to a continuing life of practice and of study of the function of the human brain. In that same year, Hendel, who had become Professor of Moral Philosophy at McGill, published a significant discussion, "The Status of Mind in Reality," in *The Journal of Philosophy* (XXXI, 9, 225–235, 1934).

In September 1973, when I had finished the first draft of *The Mystery of the Mind*, I sent it to him for criticism. In the meantime, he had returned to the United States, to become Chairman of the Department of Philosophy at Yale University, and later he had retired from that post to live in Brandon, Vermont, where he is at present engaged in the preparation for publication of his Gifford Lectures delivered at Glasgow in 1962 and 1963.

On receiving my manuscript, he replied immediately

in a long handwritten letter in which this understanding paragraph appeared:

> "As I read it [the manuscript] again and again, your story is one of your starting with the physical hypothesis (accepted by all scientists as a belief in order to gain knowledge): that the physical attributes of man and energy alone are what they can deal with. You start here and cannot do otherwise, and ought not to do so. But there are discoveries made which made you *wonder* about something that does not fit into the scientific picture, and you wonder again and again. It is the testimony of living, conscious patients. This is an *objective item* in your scientific evidence. How can it be fitted into the assumed hypothesis of an entirely physical nature of man?"

His letter led me to rewrite the monograph, making it more frankly a personal account of work and experience, a sort of pilgrim's progress. In January 1974, I sent him the final draft. I begged him then to write the Foreword to my book and hoped that he would use some of his original letter in it verbatim. In the end, what he has written will certainly make it easier for readers who are not scientists to understand my presentation, and easier to use the facts and the arguments to their own purposes.

The Search for a Mechanism of the Mind

In the past fifty years we have come to recognize an ever increasing number of semi-separable mechanisms within the human brain. They are sensory and motor. There are also mechanisms that may be called psychical, such as those of speech and of the memory of the past stream of consciousness, and mechanisms capable of automatic interpretation of present experience. There is in the brain an amazing automatic sensory and motor computer that utilizes the conditioned reflexes, and there is a highest brain-mechanism that is most closely related to that activity that men have long referred to as consciousness, or the mind, or the spirit.

Throughout my own scientific career I, like other scientists, have struggled to prove that the brain accounts for the mind. But now, perhaps, the time has come when we may profitably consider the evidence as it stands, and ask the question: *Do brain-mechanisms account for the mind?* Can the mind be explained by what is now known about the brain? If not, which is the more reasonable of the two possible hypotheses: that man's being is based on one element, or on two?

On the basis of either hypothesis the nature of the mind remains, still, a mystery that science has not solved. But it is, I believe, a mystery that science will solve some day. In that day of understanding, I predict that true prophets will rejoice, for they will discover in the scientist a long-awaited ally in the search for Truth.

Acknowledgments

Since this book is written in simple style for those in various walks of life who may be interested in an understanding of the brain and the mind of man, I have asked many whose opinion I have learned to value to read it, and I have accepted their criticisms and questions gratefully. The background of these critics varies all the way from that of my grandson, Wilder Penfield, III, who is at present a music critic, and a nephew, Mark Williamson, who is a physician skilled in computerized medicine, all the way to my contemporary in neurology, Sir Charles Symonds.

My wife, who has been the most enthusiastic and creative critic of my writing all through life, has read each portion to me, or listened to me read, and given me courage to go on in this late attempt to present the evidence so that each reader may be able to draw his or her own conclusion as to which of the two possible hypotheses about the nature of the mind seems to be the more reasonable.

The letter that came to me from Charles Symonds, (after I had sent him the manuscript) entitled Reflections, is published following the main text, as a critical epilogue, without alteration. It is a statement of his "own views on this subject" rather than an analysis of my presentation. Yet it supplements my writing admirably and gives the reader an insight into the best neurological thinking of the past century. Furthermore, his reflections give me the welcome opportunity of adding my own explanatory discussion that I have called Afterthoughts. This will, I hope, clarify the overall presentation of my thesis.

I am profoundly grateful to my son Wilder for criticism, also to my daughter, Ruth Mary Lewis, who typed the manuscript as my secretary and added many thought-provoking comments; I am grateful also to my former secretary, Anne Dawson, who edited the manuscript and to my latest secretary, Kate Esdaile, who scrutinized and marshalled the material despite my frequent changes of mind.

In conclusion, my former colleagues, Theodore Rasmussen and William Feindel have criticized this writing from the start. More than that, in their earlier years, they helped me at work and often kept me from heedless blunders.

WILDER PENFIELD
Sussex House
Austin, P. Que., Canada
August 1974

Foreword

by Charles W. Hendel, LITT.B., PH.D.*

In his Preface Dr. Penfield tells how I became associated with this piece of writing. I would like here to add something, in order to explain why, as he says, he was led "to rewrite the monograph, making it more frankly a personal account of work and experience." Earlier in my original letter I had written:

> "As I finish a second reading of your manuscript, I find the last pages are an eloquent, convincing justification of your hypothesis and belief that mind has a being distinct from body, though intimately related to and dependent on body. The final statement is you *yourself*, speaking to the reader. The careful, modest, thoughtful assertion of belief and questions alike are the marks of the philosopher. And I salute you as a master."

At the end of that letter I said:

> "For myself, reading this has been far more than instructive: it is something of an inspiration to find reasons for believing, what I have always held from the beginning of my own life and work as a philosopher: reasons for 'a persistent view in modern thought that mind is a very distinctive reality.' "

After the opening paragraph, which I have quoted above, came the fateful words. "Now I shall try to assist,

* Professor of Moral Philosophy and Metaphysics, Emeritus, Yale University.

in any way I can, to have other readers appreciate your thought and work as I do." I had in view only persuading the author to change his mode of writing, which was too much that of an address to a scientific audience, so it would reach a wider public. For it was a thrilling story to me which seemed, however, somewhat obscured by his telling how it had been designed for the meeting of the American Philosophical Society at Philadelphia, in April 1973, and then developed as a paper for the Hans Berger Centennial Symposium on the Brain-Mind Relationship in May 1973, etc. It was enlightening to read that a "suspicion about the purely physical view of mind existed in your mind quite early (Thayer Lectures in 1950 at Johns Hopkins). It was in the background. But electrical stimulation and its results made the notion develop into a suggestion or indeed even an hypothesis as you proceeded."

So what the story told was this:

> "How Wilder Penfield, operating on patients in order to cure them, if possible, found out things about the cerebral cortex and the mechanisms of the higher brain stem that turned a suspicion or mere notion into a vital hypothesis, and this further led to practical results in the lives of patients.
>
> "You became more and more convinced that the mind is something in its own right, that it did things with the mechanisms at hand in its own way, that it had an 'energy' of its own. You offer only suggestions. You discuss fairly the questions however that arise if they are taken *as hypotheses.* You end aligning yourself with the prophets, the poets, and the philosophers who have emphasized the spiritual element in man."

Having such opinions about the piece, I was moved to make the suggestion to the author himself "that you make it plainer to the reader that this piece of writing gives the autobiographical sequence of your development. Yours is a story of 'how I came to take seriously, even to believe, that the consciousness of man, the mind, is something not to be reduced to brain-mechanism.'

". . . In this development, what carries weight with the reader is the fact that in brain operations you hold a supreme position in the memory of living men and women: but *you* must only introduce the facts and experience you have recorded that disclose the ground for your belief.

"Your autobiographical material *is* powerful, the testimony of your patients is convincing, and your development toward the mystery of the mind is convincing beyond any philosopher's argument. Think it over."

The author accepted that suggestion and rewrote his monograph for the larger public than appeared intended in the original document. All that I had in mind when I spoke of a willingness "to assist" was persuading Dr. Penfield himself to take precisely that step in publication.

Then I found myself unexpectedly promoted from being a friendly critic to "philosopher" who would write a special Foreword. It seems to me that little more need be done here except to recite, as I have done, what I wrote to the author himself in the original letter. This is already accomplished in the above cited quotations, including what Dr. Penfield himself has told in the Preface.

It is possible, however, that a general public would appreciate a few remarks in addition that the author himself would not need or want.

"By the way [my letter continued] you use 'argument' here as Shakespeare does in a play. The 'argument'

> of the play is the plot. One should beware of the logicians who, when they see the term 'argument,' get off into challenges of your reasoning according to their very precise standards. Say how *you* mean 'argument' when you employ the term.
>
> "I have the same trouble with your use of 'proven' fact. Fact to us is *established* by a theoretical consideration of evidence and confirmation. Proof is not simply given with fact. The fact, as you so well know, has to be interpreted within a system of factual knowledge to be accepted as such. But don't bother about these things."

At the end of my letter was this observation that recalls our common past experience in college.

> "Your mention of James' *Psychology* touches me closely. I have never forgotten James' *Principles* and the role of the mind in purpose, attention, interest and decision. Again and again in my own experience, and study of others I have found these ideas fruitful, and indeed quite 'proven,' to my own satisfaction. They have been confirmed as being as nearly true as anything can be in regard to the nature of the mind."

Those today who have "a little philosophy" may question the author's meaning when he asks the following questions in his Preface: "Do brain-mechanisms account for the mind? Can the mind be explained by what is known about the brain?" One tends to wonder about explanation in this connection. What would be the complete explanation of the mind in terms of brain? Does it mean a "reduction" of everything mental to physical brain processes? What is an "explanation" of anything by reference to anything else?

However one answers that philosophical question it is

clear that Dr. Penfield himself is saying that the activities in which he finds evidence of mind cannot possibly be reduced to brain mechanisms—at present.

In Chapter 19 of the text, Dr. Penfield treats of the "Relationship of Mind to Brain: A Case Example." He asks himself at the end, "Has the brain explained the mind?" The question remains. One must choose, he decides, between "two explanations," which is the subject of the following chapter's discussion.

In the choice Dr. Penfield opts for a distinct element in "Man's Being," mind as well as physical brain. The reader who is familiar with the history of philosophy may call the choice that of a dualism like that of Descartes. This would be a mistake. Descartes took a position that is called "dualistic" because he discerned distinct essences, thought and extension, and found himself compelled by logic to assert both to be real and absolute, neither reducible to the other. He designated them as "substances," mind and body. But Dr. Penfield does not start with a conviction about the irreducible integrity of mind and body by themselves. This he has made clear in his Afterthoughts.

> "I do not begin with a conclusion and I do not end by making a final and unalterable one. Instead, I reconsider the present-day neurophysiological evidence on the basis of two hypotheses: (a) that man's being consists of one fundamental element, and (b) that it consists of two. . . . I conclude that there is no good evidence . . . that the brain alone can carry out the work that the mind does [and] that it is easier to rationalize man's being on the basis of two elements than on the basis of one."

Of these two basic elements of man's nature Dr. Penfield does say in Chapter 20 that "the mind must be viewed as a basic *element* in itself. One might call it a *medium*, an *essence*, a *soma*. That is to say, it has a *continuing existence*." This is the experienced character of both mind and body in the Cartesian philosophy, and it happens to be one of the features where Dr. Penfield agrees with Descartes. This is what a "substance" means in that earlier philosophy.

But this present work is not a definitive philosophy. It has emerged in a very different way, too, than did the philosophy of Descartes. Dr. Penfield welcomed the introduction to the monograph offered by his colleague Dr. William Feindel, which tells about the circumstances of the present and indicates where the work of Penfield has been heading. "Perhaps at no previous time in the history of science has there been such widespread interest, as is now evident, in the brain and its function, and how that function relates to human behavior." Dr. Feindel cites the present-day attraction of the study of the brain for "scientists trained in anatomy, physiology, pathology and other biological disciplines" as well as for "neurologists, neurosurgeons and psychiatrists," and further for those who hail from "mathematics, physics, chemistry, and electronics and computer sciences." This is the milieu in which the present study emerges.

The philosophy of the day must concern itself about the actual facts of the day's experience or it becomes irrelevant and useless. There is little profit in trying to formulate "*the* problem of mind and body" and then to relate the present "wonder" about the mind to the answer to that old question. The philosophers now must reckon with the evidence and the experiments of today and find their position accordingly.

Dr. Penfield has been "wondering" about the relationship of brain and mind: that is what we are invited to participate in here. It was Aristotle who said that philosophy begins with "wonder." We are, therefore, at the tentatives of beginning, not the conclusive ending of a philosophy.

What of the word "mystery"? Einstein is quoted: "The mystery of the world is its comprehensibility." So with the mind—we can comprehend the fact of the mystery for us. Or, since my old friend and companion of college days has referred to our working in the same class studying Immanuel Kant, I recall now what Kant said about the reality of human freedom and the reality of moral obligation in a world of scientific fact, that this is something whose very incomprehensibility in scientific terms we can, today at least, understand.

Introduction

by William Feindel, M.D., D.PHIL., F.R.C.S.(C.)*

Perhaps at no previous time in the history of science has there been such widespread interest, as is now evident, in the brain and its function, and how that function relates to human behavior. For compelling and obvious reasons, this topic has always been foremost in the attention of neurologists, neurosurgeons, and psychiatrists. Over many years, as well, the study of the brain has attracted the talents of scientists trained in anatomy, physiology, pathology, and other biological disciplines. Increasing numbers of intellectual emigrés, coming from such fields as mathematics, physics, chemistry, electronics, and computer sciences, have recently added fresh impetus to our exciting researches in the neurosciences.

It is almost a scientific cliche to say that the human brain is the most highly organized and complex structure in the universe. Made up of a dozen billion microscopic nerve-cell units interconnected by millions upon millions of conducting nerve-threads weaving incredibly intricate patterns, the brain, as an object of research, presents a defiant challenge to its own ingenuity. We have managed to work out some principles of how the brain works from careful studies of certain disorders that affect the nervous system in man. Among the most dramatic of these is epilepsy. The epileptic fit can produce movement, sensations, or changes in behavior, all quite uncontrollable by the patient: in so doing, it provides a caricature of how

* Director, Montreal Neurological Institute; William Cone Professor of Neurosurgery, McGill University.

these local nerve-cell groups might operate normally. Thus, though we still seek the exact cause of epilepsy and the reason why certain nerve cells are suddenly thrown into this violent discharge, examination and treatment of patients have led to a better understanding of the localization of function within the brain.

But there is much to consider beyond the extensive experimental work on the brain and the study at the bedside of patients with disorders of the nervous system. Today, thoughtful men continue to debate the question, as they have over the centuries, "How is brain related to mind?" Every reader will know that loss of brain produces loss of mind. But, as Sir Charles Sherrington noted, "Mind, meaning by that thoughts, memory, feelings, reasoning, and so on, is difficult to bring into a class of physical things."

Among the various groups of research workers and physicians concerned with the enormous task of exploring the "nerve-cell jungle" of the human brain, neurosurgeons alone have the unusual opportunity and privilege of being able to observe directly the living brain, and to map out its responses to stimulation, in the course of bringing therapeutic relief to their patients. And for many reasons, Dr. Wilder Penfield's distinguished contributions to this special field have been recognized as unique by his neurosurgical and scientific colleagues. During his lifelong devotion to the care of patients with focal epilepsy, he has catalogued a great body of information that has provided further insight for us into "the physiology of the mind."* He was able to do this from a substantial background of scientific preparation. As a student in the

* This phrase, Dr. Penfield notes, was used in 1872 by John Hughlings Jackson, the British neurologist who introduced the concept of increasingly complex levels of function in the brain.

laboratory of Sir Charles Sherrington at Oxford, he had the chance to develop meticulous surgical technique and systematic recording of observations, which he further refined for use in the operating room and clinic. Sherrington emphasized, on the one hand, the unity of the nerve cell and its individual processes and, on the other, the remarkable integration of these billions of units to give patterns of coordinated activity, such as the movements involved in balance, or gait, or gaze. Later, in his erudite Gifford Lectures, Sherrington brought together his physiological and historical reflections on brain and mind, to which Dr. Penfield now refers.

In another stage of scientific study, Dr. Penfield visited the Spanish histological school of Ramón y Cajal in Madrid. Cajal was then the acknowledged "maestro" of the microscopic study of the brain. He also viewed the nervous system, as did Sherrington, with the philosophic background of an expert neuroscientist. "As long as the brain is a mystery," Cajal once wrote, "the universe, the reflection of the structure of the brain, will also be a mystery."

From Cajal and his brilliant pupil del Río-Hortega, Dr. Penfield learned techniques to examine the microscopic nature of brain scars, which are sometimes associated with epilepsy. He further pursued this problem with Professor Otfried Foerster in Breslau, a neurologist turned neurosurgeon, and one of the few at that time who had persevered in the treatment of epilepsy by surgical removal of the brain scar.

Then in 1934, Dr. Penfield and his associates at McGill University established the Montreal Neurological Institute. This combined the facilities of a special hospital for nervous diseases, with the resources of brain research laboratories. Here, over a period of some thirty or more years of intense work, he directed scientific and surgical teams toward solutions of many unanswered questions about the brain, such as the mechanism of epilepsy, the learning of language, and how the brain remembers.

One example of these investigations might be given here, since it is a key point in any discussion of mind and brain. In 1952, Dr. Penfield and I were working side by side in the operating room observing the responses of patients to gentle electrical current applied to the temporal lobe. We became aware that in some patients, it was possible to produce, artificially, a curious state of automatism. During this, the patient became unaware, mouthed inappropriate comments—one patient was heard to say "time and space seem occupied"—made semi-purposeful movements, and, strangest of all, later had no memory of all this. We had come earlier to recognize, as did Hughlings Jackson years before, that similar behavior was a hallmark of a particular kind of epileptic seizure. We were now able to identify that this could be initiated from a deep part of the temporal lobe, from an amazingly local region, a small almond-shaped island of nerve cells, called the amygdala (from the Greek word for almond). It was evident that the nerve-cell discharge, set off by the stimulation, produced a train of complex events in the brain that seemed to isolate the patient's awareness and memory-recording from his motor and sensory activities. In fact, it seemed apparent that the patient had "lost his mind," in the real sense of the meaning of this ancient word.*

It was something of a minor revelation to realize that applying so slight an electrical stimulus to such a local part of the brain, could result in this profound change in

* In the English of Baeda and Chaucer, "mynd" had already taken on the sense of memory or remembering.

the patient's mental action. The immediate upshot was to indicate to us the importance of including in the surgical excision this small area affected by scar tissue, in order to eliminate the patient's seizures. This type of operation has now been done at the Institute in over 700 patients. Three out of four of these have returned to useful and happy lives, no longer under the threat of unpredictable epileptic attacks. And yet, removal of this trigger-area has not affected their minds—on the contrary, some patients, because their seizures stopped and their medicines could be reduced or eliminated, were found to function at a higher intellectual level as judged by detailed psychological tests.

This present book brings to us a distillation of the scientific writings of Dr. Penfield and his gradually evolving views of functional localization and interaction within the human brain. In addition, he has now moved on to a much deeper discussion of the brain-mind question. Philosophers and, I have no doubt, neurosurgeons, will admire his intellectual courage in tackling this project. John Locke, after mulling over his essay concerning human understanding for several years shied away from this problem, writing in his introduction, "I shall not at present meddle with the physical considerations of the mind; or trouble myself to examine wherein its essence consists. . . ." The great psychologist, William James, in whose writings Dr. Penfield found considerable inspiration, looked upon this difficulty of stating the connection between mind and brain as "the ultimate of ultimate problems."

Dr. Penfield's present analysis derives from his accumulation of direct observations on the human brain in conscious patients. In that important sense, it transcends significantly all earlier studies; either those of physiologists, who argued from a basis of experimental animal findings, or those of neurologists, psychologists, or psychiatrists, whose views were related to interpretations of the external motor and emotional behavior of patients with focal brain disorders. Many readers will recognize that the research findings summarized here by Dr. Penfield are fundamental to our understanding of memory, learning, language, and behavior. As one of his final conclusions, Dr. Penfield supports the proposition that there is something that characterizes mind as distinct from physical brain.

Many unsolved questions of the relation of brain and mind remain to be answered. As John Locke queried 300 years ago—"How do we separate imagination and madness?" We ask it even more today. What is the brain action that makes the distinction possible between these two aspects of mental behavior? What is the reason and the mechanism for the convincing kaleidoscopic activities of our dreams, which make nocturnal fools of all of us? How do we explain hypnosis, or the pain relief that sometimes comes with oriental acupuncture? What is the relation of mind to our many varieties of religious experience? And what becomes, ultimately, of that vast array of experiences, thoughts, and ideas that are entombed in our brain's memory?

The Mystery of the Mind

1 • Sherringtonian Alternatives—Two Fundamental Elements or Only One?

My professional career was shaped, I suppose, in the neurophysiological laboratory of Professor Sherrington at Oxford. Eventually it was continued in the wards and operating rooms of the Montreal Neurological Institute. Other preoccupations were many and varied, but beneath them all was the sense of wonder and a profound curiosity about the mind. My planned objective, as I turned from studying the animal brain to that of man, was to come to understand the mechanisms of the human brain and to discover whether, and perhaps how, these mechanisms account for what the mind does.

My teacher, Sir Charles Sherrington, received the Nobel Prize for his studies of reflexes and his analysis of the integrative action of the nervous system. His interest had been focused largely on the inborn reflexes, but, on retiring from the Chair of Physiology at Oxford in 1935, at the age of seventy-eight, he turned from animal experimentation to a scholarly and philosophical consideration of the brain and the mind of man.*

In the end, he could only say that "we have to regard the relation of mind to brain as still not merely unsolved, but still devoid of a basis for its very beginning." In June 1947, he wrote a foreword to his book, *The Integrative Action of the Nervous System*, which was then being republished in his honor by the Physiological Society.[32] The

* In 1937–1938, he delivered the Scottish Gifford Lectures, and published them in 1940 under the title *Man—On His Nature*.[31] (Thoughout the book, these superscript numerals refer to the numbered entries in the Bibliography, which follows the Afterthoughts.)

last paragraph of his foreword expresses his conclusion of it all:

> That our being should consist of two fundamental elements offers, I suppose, no greater inherent improbability than that it should rest on one only.

It is a quarter of a century since Sherrington wrote these words. We have learned a good deal about man since then, and it is exciting to feel, as I do, that the time has come to look at his two hypotheses, his two "improbabilities." Either brain action explains the mind, or we must deal with two elements.*

Perhaps we may take a step toward understanding, if we strive to fit each of the two hypotheses in turn to the physiological evidence that presents itself today. A good scientist is neither a monist nor a dualist while conducting his research. His chosen task is to explain everything he can by critical examination of nature and of the brain, and by planned experimentation. He will account, thus, for what he can about the universe and about man himself, having put his preconceptions out of mind. But he must stop to reconsider, too, and to rationalize from time to time.

Lord Adrian, who shared the Nobel Prize with Sherrington, spoke as a neurophysiologist in 1966 when he said: "As soon as we let ourselves contemplate our own place in the picture, we seem to be stepping outside of the boundaries of natural science." I agree with him; nevertheless, we must step across that boundary from time to

* Sherrington did not consider the third hypothesis, proposed by Bishop Berkley, that there was only one element, the mind, which explained all. The Berklian explanation assumed that matter had no existence except for its place in the mind.

time, and there is no reason to assume that critical judgment does not go with us.

Writing this book presents the author with a very exciting challenge. Accepting this, I can only give an account of my own experience, describing it simply for the clinician, the physiologist, the philosopher, and the interested layman, with apologies to each for the fact that I have not written for him separately.

A remarkable body of material has come into my hands and I have stumbled on exciting discoveries. I did summarize the material and I recorded it during and at the close of my professional career. But I turned then with great enthusiasm to authorship of another sort, perhaps unwisely. Perhaps it is one's duty to do more than make a record. In answer, I may plead that I can see it all in more mature perspective after an interval, even in the seventh and eighth decades. Is it an effort, if I may paraphrase Hamlet, to lay a "flattering unction to my soul"?

However that may be, as I turn back now to the material and reconsider a life's experience, I seem to see more clearly and understand a little better. So, I shall give the reader a brief account of this pilgrim's progress. It is a story of stumbling upon unexpected revelations, of consequent puzzlement and misconception, and of reaching higher ground to look out on thrilling new vistas of understanding. In the end I shall draw conclusions that are scientific, and present hypotheses that are obvious. After that, because these data are important in other disciplines of thought, I shall pass on to rationalization and a consideration of man's being from the point of view of a layman, and, as far as I can understand it, the point of view of philosopher and even theologian.

Can the brain explain the man? Can the brain achieve by neuronal action all that the mind accomplishes? The evidence that a clinical physiologist can gather should help to answer these questions in the end.

To see the problem of the nature of the mind more clearly, consider with me this universe of ours in long perspective. It was only after the middle millennia that life appeared—first in unicellular organisms, then gradually in more and more complicated forms, first in the sea and then on the land. It was a very recent event, as seen in this long perspective, when evidence appeared of the individual's self-awareness and purpose. Today man, with his amazing mind and his vastly complicated brain, seeks to understand the universe about him, and even the nature of life and of consciousness.

Physiologists have thrown what light they could on these things from their study of mechanisms within the body and the brain in higher and lower living organisms. They have studied sensation and movement, reflex action, and memory and behavior. Karl Lashley[7] spent thirty years of his industrious life striving to discover the nature of the "memory trace" in the animal brain, beginning with experimental investigations of the rat's brain and ending with the chimpanzee. He was hunting for the engram, the record; that is to say: "the structural impression that psychical experience leaves on protoplasm." He failed to find it and ended by laughing cynically at his own effort and by pretending to question whether, after all, it was possible for animals or even man to learn at all.

But consciousness and the relationship of mind to brain are problems difficult to study in animals. Clinical physicians, on the other hand, in their approach to man, may hope with reason to push on toward an understanding of the physiology of memory and the physical basis of the mind and of consciousness.

2 · To Consciousness the Brain Is Messenger

Hippocrates, the Father of Scientific Medicine, began to teach in the fifth century B.C. on the little Greek island of Cos. In that time, philosophers such as Empedocles and Democritus were proclaiming each his own explanation of the universe and the nature of man. Hippocrates defied what he called the "unproven hypotheses" of the philosophers, and declared that only the study and observation of nature and of man would point the way to truth.

He studied man in health and in disease, making of medicine a science and an art. But he saw in man something beyond any discovery that can be made elsewhere in nature, and thus added a moral code, a religion of medical service. In the oath that he required of his disciples there were such phrases as this: "I will use treatment to help the sick according to my ability and judgment, but never with a view to injury or wrongdoing. . . . I will keep pure and holy both my life and my art." Thus, he recognized the moral and the spiritual as well as the physical and the material.[6]

Hippocrates left behind him only a single discussion of the function of the brain and the nature of consciousness. It was included in a lecture delivered to an audience of medical men on epilepsia, the affliction that we still call epilepsy. Here is an excerpt from this lecture, this amazing flash of understanding: "Some people say that the heart is the organ with which we think and that it feels pain and anxiety. But it is not so. Men ought to know that from the brain and from the brain only arise our pleasures, joys, laughter and tears. Through it, in particular, we think, see, hear and distinguish the ugly from the beautiful, the bad from the good, the pleasant from the unpleasant. . . . To consciousness the brain is messenger." And again, he said: "The brain is the interpreter of consciousness." In another part of his discussion he remarked, simply and accurately, that epilepsy comes from the brain "when it is not normal."

Actually, his discussion constitutes the finest treatise on the brain and the mind that was to appear in medical literature until well after the discovery of electricity. It was the evidence of conduction of the brain's energy along the nerves of animals that led to the discovery of electricity itself.

In retrospect, it is abundantly clear that Hippocrates came to his conclusions by listening to epileptic patients when they told him their stories, and by watching them during epileptic seizures. The reader will come to understand, in the pages that follow, that epilepsia still has secrets to reveal. She has much to teach us if we will only listen.

Some of the notes that Hippocrates made after examining his patients were copied and recopied through the centuries. They are models of brevity and insight. Epileptic patients of a certain type, not infrequently, re-live some previous experience in which they see, perhaps, and hear what they have seen and heard at an earlier time in their lives. Realizing, as Hippocrates did, that "epilepsy comes from the brain 'when it is not normal,' " he must have guessed the truth—that the engram of experience is a structured record within the brain.*

* Although Hippocrates, because of his teaching, is to be considered the founder of biological science, his life and personality have been almost completely lost in the course of time. This led me, during the last five years of my career as an operating neuro-

It was the common understanding in those days that the soul, or consciousness, was located in the heart. For example, four hundred years later, these familiar words appeared in the Christian Gospel according to Luke: "Mary kept all these things and pondered them in her heart." When men did finally abandon the idea that thinking was carried out in the heart, and realized that the brain was the master organ, the words of Hippocrates had long been forgotten. Men thought that the brain acted somehow as a mysterious whole, sending out and receiving spirit messengers in accordance with the teaching of the physician Galen (A.D. 131–201). Long after Galen, came the discovery of animal electricity by Galvani (A.D. 1791), which banished the spirit messengers forever.

We know now that the brain does not act mysteriously, as a simple and uncomplicated whole. It has within it many partly separable mechanisms, each of them activated by the passage of electric currents along insulated

---

surgeon, to devote many of the days or weeks or months, when I could be spared from clinical responsibility, to the writing of a historical novel about the man as he must have been. It grew into a fictional presentation. I hoped, thus, that I might bring to light the real hero. (*The Torch*, Little, Brown and Co., Boston, 1960; also George Harrap & Co., London, 1961.) It was translated into some other languages, and Mr. Guram Kveladze, Chief Editor of Sabchota Sakartvelo Publishing House, who translated it into Georgian, and published it in Georgia, wrote the following to me on April 5, 1968: "The interest of Georgian readers was greater because of the fact that Hippocrates had visited Georgia and given a description of Kolhidian tribes. So the Georgian reader met in this book a highly respected and widely known personage!" Since I have had no time to follow up this interesting observation and shall not have, perhaps some Georgian scholar, seeing my note, will inform his fellow physicians in the West, how it was that Hippocrates came to cross the Black Sea, and more about his visit to Georgia.

nerve fibers. I shall point out presently that there is a specific mechanism that must be active to make consciousness and thought possible.

Can I discuss this mechanism understandably if I leave behind the technical phrase and speak the language of the unspecialized but educated man? I dare say Benjamin Franklin, founder of the American Philosophical Society, explained in easily understood excitement to the first members of that society how it is that electricity passes down the wet string of a kite. I wish I had been there. Perhaps I would have understood the nature of this all-important wonder called electricity. It seems to me that, somehow, it is like the mind in the sense that one cannot assign to the mind a position in space and yet it is easy to see what it does and where it does it.*

* Hans Berger, the discoverer of electroencephalography, when he hoped (vainly) to record the activity of the mind electrically, may have had in mind this similarity.

3 · Neuronal Action within the Brain

Definitions are useful at the beginning of an essay—although the text, in this case, will certainly show them to be inadequate. The *mind* (or spirit) is, to quote from Webster's Dictionary: "the element . . . in an individual that feels, perceives, thinks, wills, and especially reasons."

The *brain* is the vastly complicated master organ within the body that makes thought and consciousness possible. In its integrative and coordinating action, it resembles in many ways an electrical computer. An individual *brain-mechanism* is a functional unit that plays a somewhat specialized role in the total integrative action of the brain.

Each nerve cell, or *neurone*, is capable of developing its own electrical charge. Each has one branch called the axone, among its many branches. The axone carries a current of neuronal impulses outward, away from the cell, to other cells. The arriving impulses stimulate each target cell to flash the message onward, or they check activity in the target cell producing inhibition of cell activity.

The cell bodies are collected together, forming islands, or blankets of *gray matter*. The branching connections form the *white matter*. This whole system vibrates, one might say, with an energy that is normally held in disciplined control, like that of a vast symphony orchestra, while millions of messages flash back and forth, to as many functional targets.

However, when some abnormality presents itself within the skull, and becomes a chronic abnormality that irritates the gray matter, it forms a focus of irritation and may cause a recurring disorderly explosion of energy involving many cells at once, like lightning from a miniature thunder cloud. Each time this happens, an *epileptic fit* comes to some unfortunate victim. The attack varies in outward character according to the function of the gray matter in which the discharge takes place. If it occurs in the cells of the gray matter that forms a part of one of the sensory circuits, a sensation is felt; if it occurs in cells of the motor system, movement follows. *Epilepsy*, which is the name for the tendency to these attacks, is as old as the history of man. Indeed, it is probably much older, since it attacks animals far more primitive than man.

When the Montreal Neurological Institute opened its doors in 1934, we had available, at last, facilities for studying the human brain as well as for treating its disabilities. I had learned to operate on epileptic patients like those who taught Hippocrates so much. In some cases, we could remove the cause or remove the altered portion of brain in which the epilepsy-producing discharge began.

Our purpose, of course, was always to cure. And the patient, who remained awake and alert through long operations, carried out in the hope of cure, did guide the surgeon's hand. More than that, the patient taught us much in the process.*

* These operations could be done safely, and with a reasonable chance of cure, only when the surface of one hemisphere of the brain was exposed widely for careful study and possible excision. There was less danger to life and a better chance to understand each patient's problem if consciousness could be preserved throughout the procedure. Local analgesic was therefore injected into the scalp to prevent pain, and no sedative or anesthetic was given. To be successful, as well as humane, it was

Since a gentle electric current interferes with the patient's use of a convolution of the brain and sometimes produces involuntary expression of its function, a stimulating electrode could be used to map out the cortex and to identify the convolutions as the patient described his sensations and thoughts. Also, the electrode, if used with discretion, would sometimes reproduce the beginning of the patient's epileptic seizure and, thus, disclose the site of brain irritation. By talking to the patient and by listening to what came into his mind each time the electrode was applied to the cortex, we stumbled upon new knowledge. If we removed convolutions as treatment for the fits, we learned about brain function in another way as soon as the nature of the patient's loss was determined after the operation.

The observations to be presented in the following discussions, Chapters 4 through 9, have been published from time to time with the help of a succession of able associates in the Montreal Neurological Institute, not all of whom are named in the bibliography.

---

essential for the surgeon to explain each step. Indeed, he must take time for talk before and during the operation. He must, in fact, be the patient's trusted friend.

4 · Sensory and Voluntary-Motor Organization

Here, then, is a brief outline of the sensory and motor mechanisms and of some of the inborn reflexes that play roles in the *integrative action* of the brain of man and other mammals. I hope it may serve as a preparation for some readers and a review or revision for others, before I pass on to the discussion of brain-mechanisms that are more closely related to the action of the mind.

The brain-stem and the spinal cord provide man with inborn reflexes, as they do the other mammals. They regulate such things as muscle tone, maintenance of posture, mechanics of walking, temperature control and sleep-rhythm, breathing, and coughing (see Figure 1).

The cerebral hemispheres that make up the telencephalon, or new brain, grow out of the diencephalon, which may be called the higher brain-stem or old brain. The hemispheres increase in proportional size from the lower vertebrates on up to man. The inflow of nerve impulses carrying pain sensation, for example, passes inwards and upwards through the spinal cord and lower brain-stem to a nucleus of gray matter within the diencephalon. This is the target-gray matter for pain. Pain differs from other forms of sensation since it makes no detour to the cerebral cortex. But the other bundles of fibers carrying sensory impulses that will be converted into discriminatory sensation make important detours. These streams provide information for appreciation of touch, position, vision, hearing, taste, and smell. Each stream comes to a first cellular interruption in the gray matter within the higher brain-stem, but continues on (with the possible exception of smell) in a detour out to a second cellular inter-

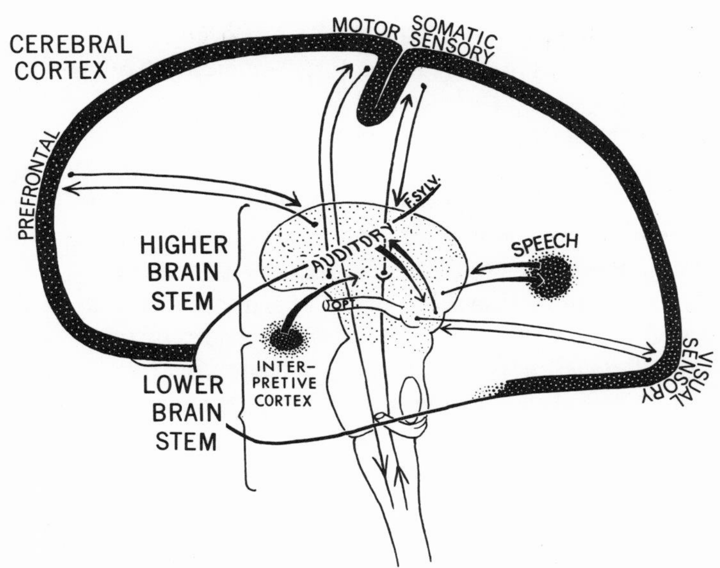

FIGURE 1. *Some Brain-Mechanisms.*

The cortex of the left hemisphere of the human brain is shown in stippled black, the brain-stem and spinal cord are shown in outline within. The principal direction of flow of electrical potentials to and from subdivisions of the cortical gray matter is indicated by the arrows in certain mechanisms as follows: *Motor*—from higher brain-stem to motor cortex and on down to motor cells in the lower brain-stem or spinal cord, producing voluntary movement; *Somatic Sensory*—from eye, ear, body, and limbs upward to higher brain-stem, then in a detour out to somatic sensory convolution and back to higher brain-stem; *Visual Sensory*—from retina through brain-stem (optic thalamus) to visual sensory convolutions of the cortex and back to brain-stem; *Auditory*—from inner ear through brain-stem (medial geniculate body) to Heschl's auditory convolutions and back to the higher brain-stem; *Speech*—from higher brain-stem to speech cortex and back again; *Prefrontal*—from higher brain-stem to prefrontal cortex and back; *Interpretive*—the arrow indicates one part of a circuit yet to be fully demonstrated. This part, as proven by electrode stimulation of interpretive cortex, activates gray matter apparently located in the higher brain-stem. The result is a "flashback" from the record of past experience.

In general, the cerebral cortex seems to play a role in the elaboration of function in each mechanism. The higher brain-stem initiates activity in the mechanism or receives the flow of electrical potentials for further integrative action.

ruption in the gray matter of the cerebral cortex. From there it returns directly to the target-nucleus of cells within the gray matter of the higher brain-stem.

Man's auditory cortex (Heschl's gyrus within the fissure of Sylvius, as shown in Figures 1 and 8) is committed to serve the purposes of auditory sensation. The stream of neuronal information from the ear comes to the higher brain-stem and detours out to Heschl's gyrus. After a cellular junction in the gray matter of that convolution, it flows back into the higher brain-stem. The same is true of the visual sensory cortex, as can be seen in Figure 1. It is a way station between the eye and higher brain-stem.

In this brief outline of the afferent sensory circuits, I have made no reference to the reticular formation described in the brain-stem by Moruzzi and Magoun. Time will doubtless show the functional importance of this system during centrencephalic integration.[8,9]

Recent studies show that each sensory input, whether auditory or visual, or from the great somatic sensory systems of the body, gives off collateral branches on its way to the thalamus, the uppermost nucleus in the brain-stem. These collaterals feed into the reticular formation of the brain-stem. This may well give the reticular formation a means of inhibiting or reinforcing incoming sensory messages in relation to the thalamic or cortical reception of those messages.

This is all part of the centrencephalic system of functional integration that makes possible sensory-motor reaction, as well as conscious reaction and planned action.

In general, it is clear that all sensory data that could inform the individual about his environment are conducted by afferent streams of electrical potentials to gray matter of the higher brain-stem, directly or indirectly.

The word *afferent* means a carrying toward an objective. *Efferent* is to carry away from a source. When considering the functional organization of the brain, afferent suggests movement toward gray matter in the higher brain-stem.

On the other hand, the stream of nerve impulses that controls voluntary activity is efferent. It passes from gray matter in the higher brain-stem outward, making its own detour out to the motor convolutions of the cerebral cortex. After a cellular break there, it passes directly back to the lower brain-stem and spinal cord for a final cellular break before it reaches the muscles. This motor outflow directs activity that may be planned or voluntary.

The sensory and the motor convolutions in man and other mammals are committed as to function at birth. The hippocampal zone (see Figure 8), on the undersurface of each temporal lobe, is likewise committed to its function. It plays a certain role in scanning the record of past experience and in memory recall. On the other hand, some of the convolutions that are used eventually for what may be called psychical functions are uncommitted to their exact function at the time of birth as well, as will be explained presently.

5 • The Indispensable Substratum of Consciousness

Gradually it became quite clear in neurosurgical experience, that even large removals of the cerebral cortex could be carried out without abolishing consciousness. On the other hand, injury or interference with function in the higher brain-stem, even in small areas, would abolish consciousness completely.

An invitation to give the Harvey Lecture at the New York Academy of Medicine caused me to review and reconsider functional localization in 1938.[14] In summary, this was the conclusion:

> There is much evidence of a level of integration within the central nervous system that is (functionally) higher than that to be found in the cerebral cortex, evidence of a regional localization of the neuronal mechanism involved in integration. I suggest that this region lies not in the new brain (the cortex) but in the old (the brain-stem).

And again this: "The *indispensable substratum* of consciousness lies outside the cerebral cortex, probably in the diencephalon (the higher brain-stem)." The realization that the cerebral cortex, instead of being the "top," the "highest level" of integration, was an elaboration level, divided sharply into areas for distinct functions (sensory, motor, or psychical), came to me like a bracing wind. It blew the clouds away and I saw certain *brain-mechanisms* begin to emerge more clearly, and they included those of the mind.[15,16]

Somewhat later I realized that man has convolutions that are new from the point of view of evolution, and not committed to motor or sensory function. They are to be programmed as to function after birth. As compared with other mammals, man has a very considerable enlargement of the cerebral hemispheres in two major areas: a) prefrontal, and b) temporal, as shown in Figure 2. Both additions have to do with what one may call the transactions of the mind.

a) One may surmise something about the function of the first-mentioned addition if it is stated that a major removal of the anterior portion of the frontal lobe results in a defect in the patient's "capacity for planned initiative" (Penfield and Evans,[23] 1940).

b) The second addition enlarges man's temporal lobes. New convolutions appear there between the auditory sensory-cortex and the visual sensory-cortex, crowding those two sensory areas right off the surface in each hemisphere and into the depth of the fissures, and forming a temporal pole in front and below.

When a child is born, the new convolutions of the temporal lobe are uncommitted and unconditioned as far as function is concerned. During the initial learning period of childhood, some of these convolutions will be programmed for *speech* on one side or the other, usually the left side in right-handed individuals. The rest of them will be devoted to interpretation of present experience in the light of past experience. This we have labeled the *interpretive cortex.* These new areas of cerebral cortex, both frontal and temporal, are employed in the mechanisms of mind-action after the early period of what may be called conditioning or programming. This is explained in the next chapters.

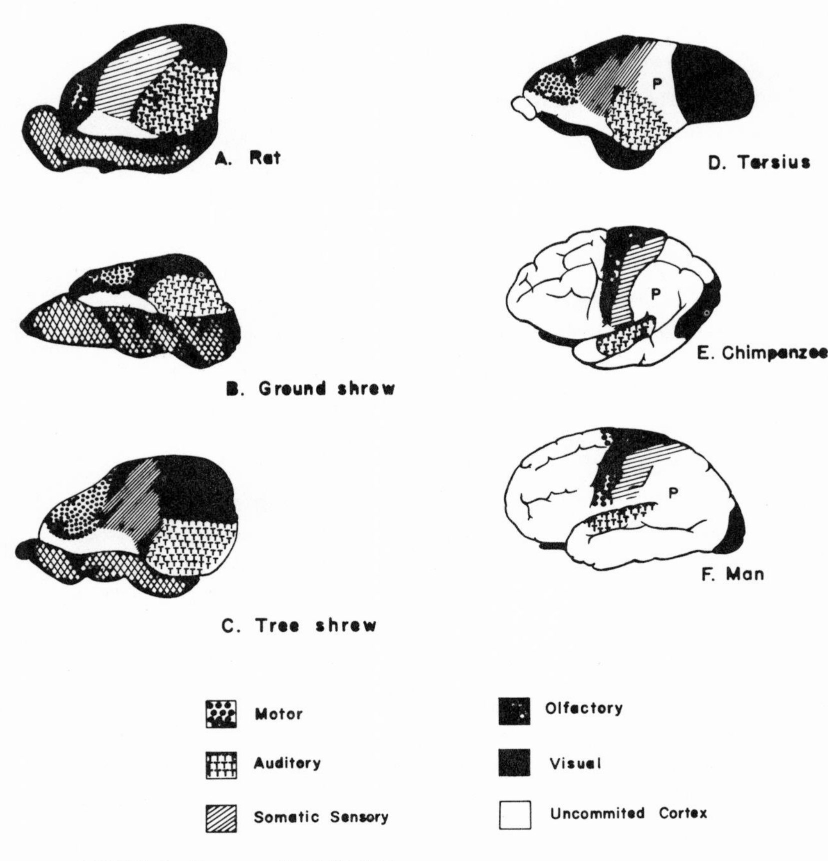

FIGURE 2. *Uncommitted Cortex.*

Functional diagrams of the cerebral cortex of some mammals. The blank areas suggest the approximate extent of gray matter that is not committed to motor or sensory function at birth. In man, for example, the auditory sensory-cortex has really been crowded off the external surface of the brain into the fissure of Sylvius. For this figure, I am indebted to the late Stanley Cobb.

6 · The Stream of Consciousness Electrically Reactivated

In the course of surgical treatment of patients suffering from temporal lobe seizures (epileptic seizures that are caused by a discharge that originates in that lobe), we stumbled upon the fact that electrical stimulation of the interpretive areas of the cortex occasionally produces what Hughlings Jackson had called "dreamy states," or "psychical seizures" (Jackson[3,4]). Sometimes the patient informed us that we had produced one of his "dreamy states" and we accepted this as evidence that we were close to the cause of his seizures.* It was evident at once that these were not dreams. They were electrical activations of the sequential record of consciousness, a record that had been laid down during the patient's earlier experience. The patient "re-lived" all that he had been aware of in that earlier period of time as in a moving-picture "flashback."

On the first occasion, when one of these "flashbacks" was reported to me by a conscious patient (1933), I was incredulous. On each subsequent occasion, I marvelled. For example, when a mother told me she was suddenly aware, as my electrode touched the cortex, of being in her kitchen listening to the voice of her little boy who

* Electrical exploration was a particularly helpful guide to our surgical procedures before the development of electroencephalography and electrocorticography. Herbert Jasper came to the Montreal Neurological Institute in 1935, bringing with him this new electrographic technique and his invaluable neurophysiological collaboration. This constructive cooperation was to result in a book, *Epilepsy and the Functional Anatomy of the Human Brain*, in 1954.[25]

was playing outside in the yard. She was aware of the neighborhood noises, such as passing motor cars, that might mean danger to him.

A young man stated he was sitting at a baseball game in a small town and watching a little boy crawl under the fence to join the audience. Another was in a concert hall listening to music. "An orchestration," he explained. He could hear the different instruments. All these were unimportant events, but recalled with complete detail.

D.F. could hear instruments playing a melody. I re-stimulated the same point thirty times (!) trying to mislead her, and dictated each response to a stenographer. Each time I re-stimulated, she heard the melody again. It began at the same place and went on from chorus to verse. When she hummed an accompaniment to the music, the tempo was what would have been expected.

In other cases, different "flashbacks" might be produced from successive stimulations of the same point. Perhaps it may add realism if I describe here one illustrative case briefly, although it has been published already.[18] For the sake of those who are not clinicians, I shall even include a photograph of the patient in position for operation.

M.M., a young woman of twenty-six (Figure 3), had minor attacks that began with a sense of familiarity followed by a sense of fear and then by "a little dream" of some previous experience. When the right hemisphere was exposed at operation, as shown in Figure 4, I explored the cerebral cortex with an electrode, placing numbered squares of paper on the surface of the brain to show the position each time a positive response was obtained. At point 2 she felt a tingling in the left thumb; at point 3, tingling in the left side of the tongue; at 7 there was movement of the tongue. It was clear then that

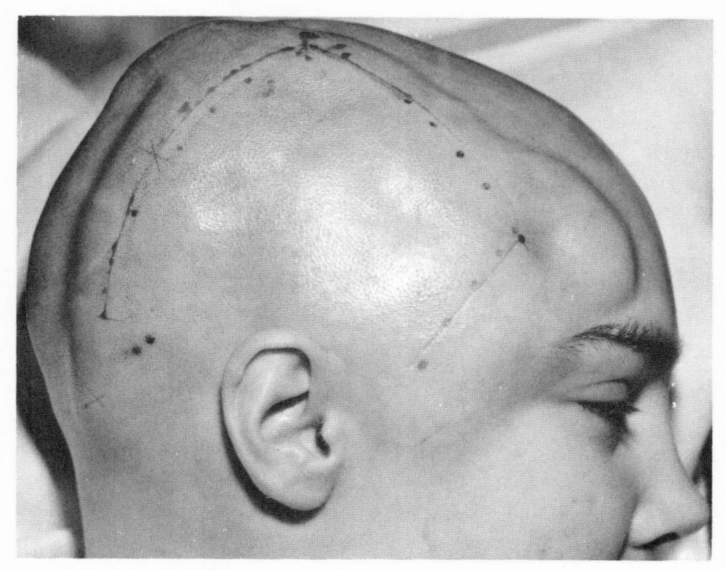

FIGURE 3. *Case M.M.*

The patient is lying on the operating table. A local analgesic has been injected into the scalp and the incision has been marked out by scratches on the skin. See Bibliography.[19]

This photograph is introduced to remind the reader that although surgeon and patient are hidden from each other by a sterile sheet during operation, they are near each other. Sympathy and mutual understanding have helped these patients to discuss their thoughts and feelings freely during electrical stimulation of the brain and removal of scarred convolutions. Although the brain is not sensitive in itself, and cannot give rise to pain, the operations are sometimes long and dangerous and very tiring. The intelligent interest and accurate reporting of these invariably gallant friends has contributed greatly to this study of the physiology of the mind.

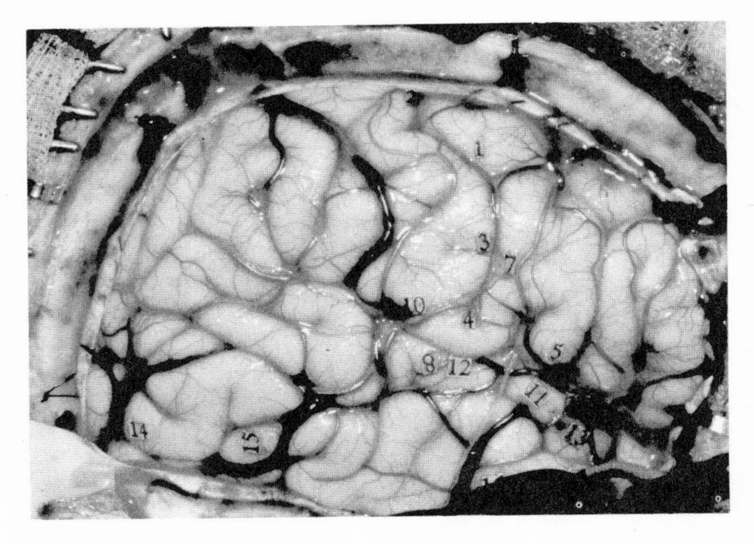

FIGURE 4. *Case M.M.*

Right hemisphere exposed. The numbered tickets mark points where there were responses to the surgeon's stimulating electrode.

3 had been placed on the somatic sensory convolution and 7 on the motor convolution (Figure 5). It is now obvious that 11 marks the first temporal convolution below the fissure of Sylvius. My postoperative sketch, seen in Figure 5, shows the position of all the stimulation points that gave rise to positive responses. The stimulating current was increased from two to three volts. The succeeding responses from the temporal lobe were "psychical" instead of sensory or motor. They were activations of the stream of consciousness from the past as follows:

> 11—"I heard something, I do not know what it was."
>
> 11—(repeated without warning) "Yes, Sir, I think I heard a mother calling her little boy somewhere. It seemed to be something that happened years ago." When asked to explain, she said, "It was somebody in

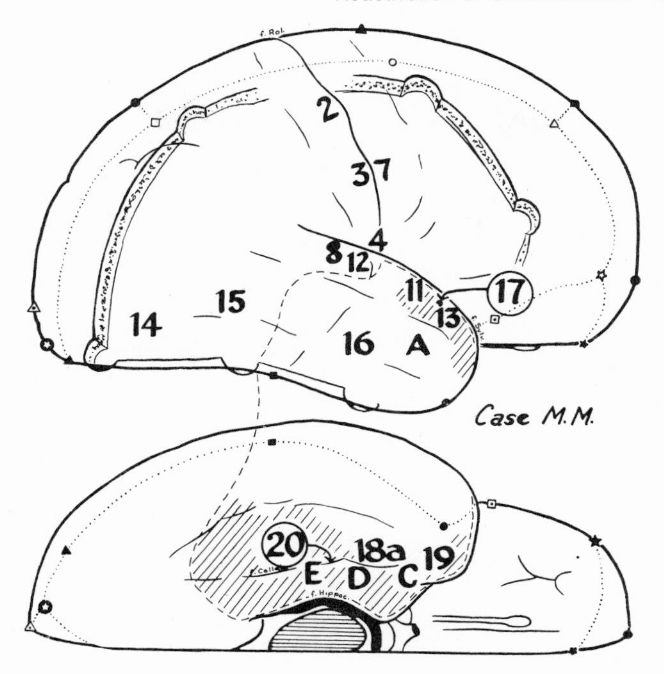

FIGURE 5. *Case M.M.*

Diagram of the operative field and the points of positive response. The broken line shows the extent of removal of the temporal lobe in treatment of the focal epilepsy. Shading indicates the area of sclerosis and atrophy due, in all probability, to pressure upon the brain at the time of birth.

the neighborhood where I live.'' Then she added that she herself ''was somewhere close enough to hear.''

12—''Yes. I heard voices down along the river somewhere—a man's voice and a woman's voice calling . . . I think I saw the river.''

15—''Just a tiny flash of a feeling of familiarity and a feeling that I knew everything that was going to happen in the near future.''

17c—(a needle insulated except at the tip was

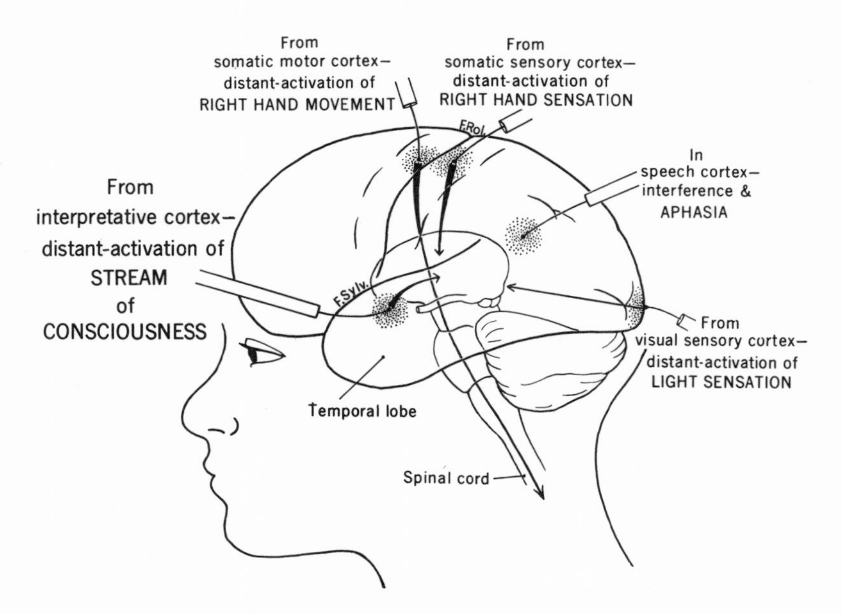

FIGURE 6. *Activation of the Brain's Record of Consciousness and Some Other Results of Stimulation.*

The left hemisphere of the brain is outlined, with the brain-stem and spinal cord shown beneath, to illustrate the results of electrode stimulation of the cortex in motor, sensory, and what may be called psychical areas for the recall of past experience. The dotted zone about each stimulating electrode tip (unipolar) suggests the area of interference in which local cortical elaborative action is arrested by electrical interference. In addition to this interference, a positive response is described from each of these electrodes, except the one on the area where speech is localized. Stimulation of the speech cortex produces only interference aphasia. The positive responses, on the other hand, are caused by normal axonal conduction from cells near the electrode to a distant but functionally related area of gray matter. Thus, the active response is a physiological activation of that distant gray matter. In the case of stimulation of the interpretive cortex, it is the sequential record of successive conscious states from the past that is activated. In the case of motor cortex the target of activation is gray matter in lower brain-stem or spinal cord. In the case of sensory areas the target is in the higher brain-stem.

inserted to the superior surface of the temporal lobe, deep in the fissure of Sylvius, and the current was switched on) "Oh! I had the same very, very familiar memory, in an office somewhere. I could see the desks. I was there and someone was calling to me, a man leaning on a desk with a pencil in his hand."

I warned her I was going to stimulate, but I did not do so. "Nothing."

18a—(stimulation without warning) "I had a little memory—a scene in a play—they were talking and I could see it—I was just seeing it in my memory."

I was more astonished, each time my electrode brought forth such a response. How could it be? This had to do with the mind! I called such responses "experiential" and waited for more evidence. Meanwhile, through the early years, we were very busy charting out the sensory, the motor, and the speech areas of the human cortex.[29,30]*

* The first of the books referred to, *The Cerebral Cortex of Man*,[29] published in 1950, was the result of a collaborative review of all our evidence in the Montreal Neurological Institute up to that date drawn, for the most part, from mapping studies of the cortex with a stimulating electrode. The evidence was amplified by careful excisions of convolutions in the treatment of epilepsy. Theodore Rasmussen, as Director of the Montreal Neurological Institute, continued these studies of the sensory and motor cortex and continues to do so now that he has retired as Director in favor of William Feindel.

In the production of the book *Speech and Brain-Mechanisms*[30] in 1959, my collaboration with Lamar Roberts continued over a ten-year period. It showed that in the adult the speech mechanism consists in three cortical areas (temporal, inferior frontal and mid-frontal), coordinated by one gray matter center in a thalamic nucleus of the higher brain-stem. Integration of this amazingly well-localized speech mechanism, with the *highest brain-mechanism* and the *automatic sensory and motor mechanism*, as will be shown in subsequent chapters, is carried out by a *centrencephalic coordinating system.*

7 · Physiological Interpretation of an Epileptic Seizure

In 1958, after I had accumulated considerable clinical experience, I reconsidered critically the physiology involved in the electrical exploration of the human brain. This was reported in the Sherrington Lecture.[18] I realized that when an electrode passes a current into the cerebral cortex, the current interferes completely with the patient's normal use of that area of gray matter. In some areas, there is no evidence of any further effect. For example, as shown in Figure 6, an electrode on one of the three areas of speech cortex causes aphasia. But, in other areas, as explained in Table I, stimulation gives a positive re-

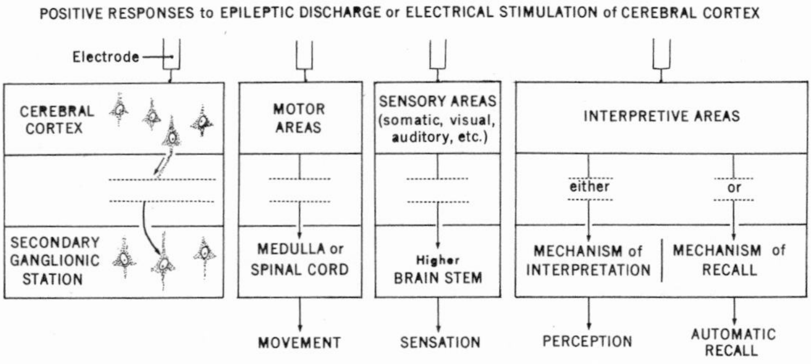

POSITIVE RESPONSES to EPILEPTIC DISCHARGE or ELECTRICAL STIMULATION of CEREBRAL CORTEX

Electrode

CEREBRAL CORTEX

SECONDARY GANGLIONIC STATION

MOTOR AREAS

MEDULLA or SPINAL CORD

MOVEMENT

SENSORY AREAS (somatic, visual, auditory, etc.)

Higher BRAIN STEM

SENSATION

INTERPRETIVE AREAS

either

or

MECHANISM of INTERPRETATION

MECHANISM of RECALL

PERCEPTION

AUTOMATIC RECALL

TABLE I. *Positive Responses.*

Electrical stimulation (or epileptic discharge) interferes with function of gray matter locally. It produces an active response only when the electrode is applied to an area of cerebral cortex from which axonal conduction along a functional tract normally activates some *distant* ganglionic station. Cortical responses are of four types: muscular movement, sensation, interpretive perception, and recall of conscious experience.

sponse as well. Such positive responses are produced not by activation of the local gray matter near the electrode, but by neuronal conduction along insulated axones to a distant area of gray matter that is beyond the interfering influence of the electrode's current.

Let me repeat: The activation is of the distant gray matter. See in Figure 6 stimulations of motor cortex, also, somatic sensory-cortex and visual sensory-cortex. There is always interference in the normal use of the local gray matter. If there is also a positive response, it is due to functional activation of distant gray matter. Consequently, when the electrode is applied to the hand area of the motor cortex, the delicate movements of the hand, which the cortex makes possible, are paralyzed, but the secondary station of gray matter in the spinal cord is activated, and crude movements, such as clutching, movements of which an infant is capable, are carried out.

In clinical epilepsy, the spontaneous discharge occurs, in the great majority of cases, either in the gray matter of the cortex or in the gray nuclei of the higher brain-stem. It never occurs in white matter. If it occurs in a so-called silent area of the cortex, there may be no manifestation of it unless an electroencephalogram is being taken.

In any fit, focal discharge begins in some local region of gray matter. If a positive manifestation occurs, it is produced, as in the case of electrical stimulation (see Table I), by axone-conduction to a distance. It is due, then, to neuronal activation of some distant secondary ganglionic station (see Figure 6). An epileptic discharge continues until the discharging local neurones are exhausted. The secondary distant response, which it produced, also stops then, but the local paralytic interference in the primary area of discharge continues after the discharge is over until there is recovery from the cell exhaustion. The distant response, if any, is a physiological phenomenon and stops, as I have said, as soon as axonal conduction to it stops.

There is always a danger, in electrical exploration, that stimulation by the electrode may bring to the cortex a current that is too strong. The local gray matter then goes into epileptic discharge. When the electrode is withdrawn there is an after-discharge and a local seizure. There is also added danger then that axonal conduction from the local gray to some distant gray matter may have increased enough to become a bombardment, and so produce a secondary epileptic explosion.

Spread of the local discharge in any fit may occur in one of two ways: (1) by a "Jacksonian march" into contiguous gray matter, or (2) at a distance (as just explained) by neuronal conduction to a functionally related area of gray matter. Spread of discharge, and, thus, of the epileptic fit, occurs when that conduction turns into a too violent bombardment. The physiological activation of the distant gray matter is then replaced by discharge in that distant area. That causes a new local functional interference at a distance instead of activation.*

* If this is a true statement of the physiological principles involved in epileptic seizures, and I believe it is, it calls for the thoughtful attention of clinicians and electroencephalographers.

8 · An Early Conception of Memory Mechanisms — And a Late Conclusion

This late understanding of the physiology of electrical stimulation, and of the pattern of neuronal discharge in an epileptic seizure, led at once to a clearer understanding of what is taking place in each *experiential response* to electrical stimulation. It called for a reconsideration of the "flashbacks." Consequently, after the close of my own career as an operating neurosurgeon in 1960, we reconsidered and published every detail of the experiential responses so others might judge their meaning for themselves. These were presented by Penfield in the Lister Oration, Royal College of Surgeons, in 1961, and published in full with Phanor Perot in 1963.[28]

There were 1,132 patients for us to reconsider. The brain of each had been explored under local anesthesia in the course of an operation for radical treatment of epilepsy. In 520, the temporal lobe was exposed and explored. The experiential responses came only from the temporal lobe, never from any other part of the brain. Of the temporal explorations 40, or 7.7 percent, gave experiential responses; 53 patients, or 10 percent, had complained of dreamlike attacks, in which past memories came to mind, before operation.

In 1951,[16] I had proposed that certain parts of the temporal cortex should be called "memory cortex," and suggested that the neuronal record was located there in the cortex near the points at which the stimulating electrode may call forth an experiential response. This was a mistake, as shown clearly in 1958 during my Sherrington Lecture.[18] The record is *not* in the cortex. Nevertheless, the initial hypothesis proposed at that time is still tenable: "It is tempting to believe," I wrote, "that a synaptic facilitation is established by each original experience." If so, that permanent facilitation could guide a subsequent stream of neuronal impulses activated by the electric current of the electrode even years later.

Since then, as I have already pointed out, we have come to call the "memory cortex" by another name—the "interpretive cortex." Its boundaries and those of the major speech area may be seen in Figures 7 and 8. And today, we realize that stimulation of the interpretive cortex activates a record located at a distance from that

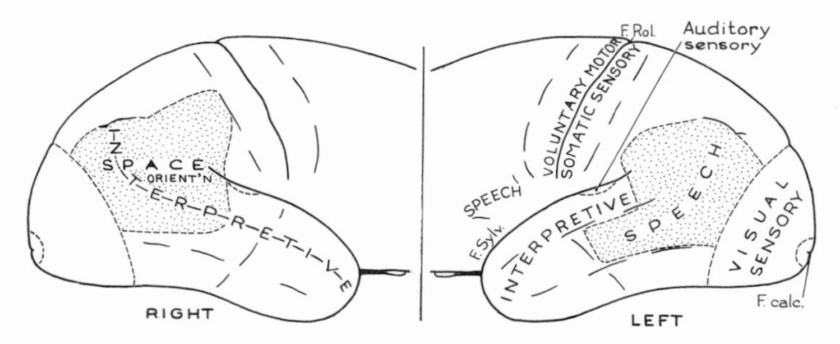

FIGURE 7. *Interpretive Cortex and Speech Cortex (see also Figure 8).*

Lateral surfaces of the posterior parts of both hemispheres of a human adult. On the dominant, or speech side, interference aphasia is produced by stimulating in the area marked *speech.* Both experiential and interpretive responses are produced by stimulating in the interpretive cortex. The area marked *space orientation* on the non-dominant side (right) was outlined by study of the results of cortical excision. Complete removal of this area produced permanent spatial disorientation without aphasia.*

* This figure is from Penfield.[22] For evidence in regard to the frontiers of the temporal speech area (Wernicke), and also for that on space orientation, see Penfield and Roberts.[30] For the localization of the interpretive cortex, see Penfield.[19]

cortex, in a secondary center of gray matter. Putting this together with other evidence makes it altogether likely that the activated gray matter is in the diencephalon (higher brain-stem), as I shall describe below.

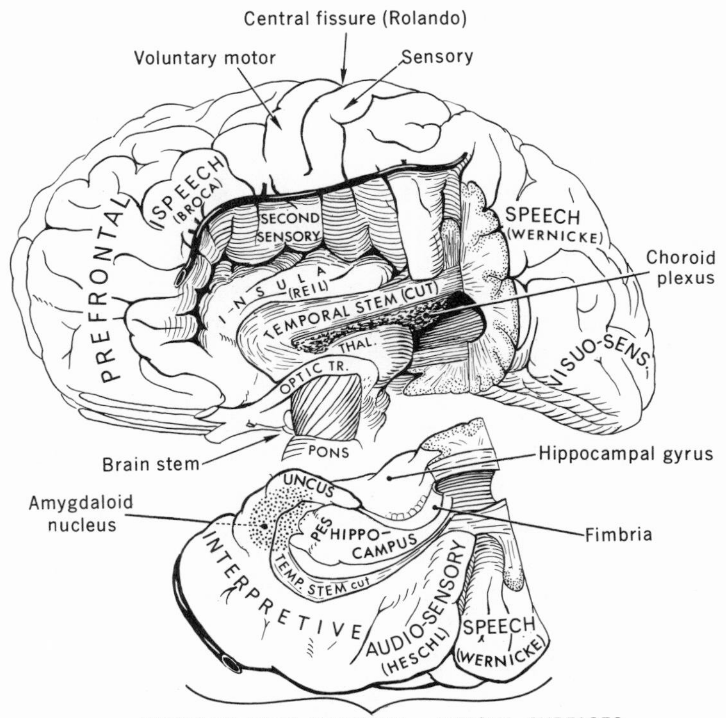

FIGURE 8.

Left cerebral hemisphere. The temporal lobe was dissected free at autopsy by opening the fissure of Sylvius. It was then cut across and turned down. Note that the hidden audiosensory gyrus of Heschl is seen to be bounded by speech cortex posteriorly and interpretive cortex anteriorly. (This drawing, like most of the preceding ones, was made by Miss Eleanor Sweezey.)

9 · The Interpretive Cortex

Let me marshal and reconsider the evidence now presented to us by epilepsy and the electrode, after which we may go on to a consideration of the relationship of mind to brain, in Chapters 10 to 16.*

Two related mechanisms are revealed by stimulation of the interpretive cortex (Figures 7 and 8). Each of them was activated in Case M.M. above:

(a) There is a brain mechanism, the function of which is to send neuronal signals that interpret the relationship of the individual to his immediate environment. The action is automatic and subconscious, but the signal appears in consciousness. Such signals as this: these things are "familiar" or "frightening." They are "coming nearer" or "going away," and so on.[10,19]

(b) Secondly, there is another, related, brain mechanism that is revealed in experiential responses like those described in the case of M.M. (Chapter 6) and the others. The mechanism is capable of bringing back a strip of past experience in complete detail without any of the fanciful elaborations that occur in a man's dreaming.[19,28] In ordinary life, the automatic signal that informs one that present experience is familiar comes to all of us, I suppose. If it is accurate, and it usually is, one must be

* I began to do this in a chapter of the book entitled *Basic Mechanisms of the Epilepsies* (edited by my former associates, Herbert Jasper, Arthur Ward, and Alfred Pope).[22] It led me to a discussion of a specific mechanism for the mind. I shall now push the argument through to a conclusion.

using an automatic mechanism that can scan a record of the past, a record that has not faded but seems to remain as vivid as when the record was made.*

The gray matter of the interpretive-cortex is part of a mechanism that presents interpretations of present experience to consciousness. In a sense, it would seem that the interpretive-cortex does for perception of non-verbal concepts what the speech-cortex and the speech-mechanism do for speech. The localization of areas devoted to speech is reasonably clear. Although much work has yet to be done on the recognition of non-verbal concepts, I shall refer to the mechanism now as the non-verbal concept mechanism. These mechanisms, the one verbal and the other non-verbal, form a remarkable memory file to be opened either by a conscious call or by an automatic one.[18,19,20]

There is much more to be said about the temporal lobes and memory when time permits. That mysterious doubled structure, the hippocampus, may well have much to do with memory of smell in some lower mammals, but in man, it is concerned with memory of other things. It can be removed on one side with impunity when the remaining hippocampus is functioning normally. But, if it is removed on both sides, the ability to reactivate the record of the stream of consciousness, voluntarily or

* Although the great majority of experiences thus recalled has been strongly visual or strongly auditory, or both, the perception of familiarity is not limited to auditory or visual experience at all, but apparently applies to all that enters consciousness. A person seen may be labeled as "seen before" (*déjà vu*), a bar of music as "heard before," a sequence of events as "happened before."

automatically, is lost. The hippocampi seem to store keys-of-access to the record of the stream of consciousness. With the interpretive cortex, they make possible the scanning and the recall of experiential memory. See Penfield and Mathieson.[27]

10 · An Automatic Sensory-Motor Mechanism

And now there opens before us an exciting vista in which the automatic mechanisms of the brain interact with, and may be separated from, the brain's machinery-for-the-mind.

As I have pointed out, epileptic discharge may, and frequently does, confine itself selectively to one functional system, one functional mechanism within the brain. When it does so, it paralyzes that mechanism for any normal function. If the function of gray matter is highly complicated and only partially automatic, such as in the speech area of the human cerebral cortex, the epileptic discharge in it produces nothing more than paralytic silence, e.g., aphasia.

And so it is that the mechanism in the higher brain-stem, whose action is indispensable to the very existence of consciousness, can be put out of action selectively! This converts the individual into a *mindless automaton*. It happens when epileptic discharge occurs in gray matter that forms an integral part of that mechanism. The tentative localization of that gray matter is shown in Figure 9. If the discharge occurs there, primarily, the patient's attack is called *petit mal* automatism. But as I have already pointed out, the temporal cortex and the prefrontal cortex have much to do with the transactions of the mind, and a seizure discharge, which begins locally in temporal cortex or in the anterior frontal cortex, may spread by violent distant bombardment to this gray matter in the higher brain-stem and, thus, produce an attack of autom-

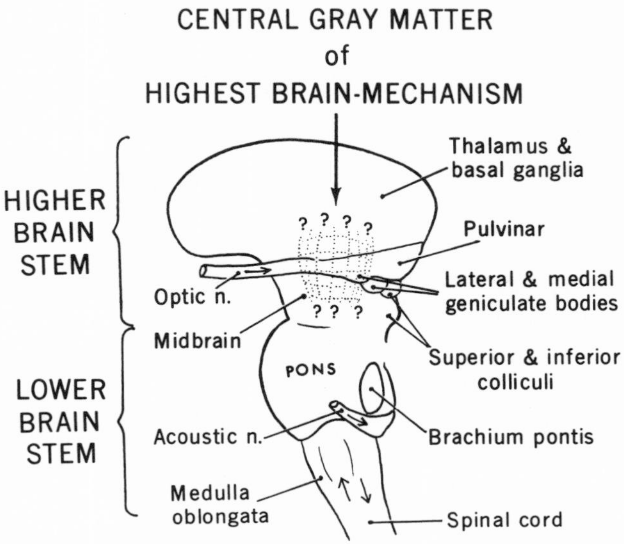

FIGURE 9. *The Highest Brain-Mechanism.*

The site of the central gray matter of this brain-mechanism, the normal action of which constitutes the physical basis of the mind, is shown by the dotted lines. The question marks indicate only that the detailed anatomical circuits involved are yet to be established, not that there is any doubt about the general position of this area in which cellular inactivation produces unconsciousness. Such inactivation may be brought about variously by pressure, trauma, hemorrhage, and local epileptic discharge; it occurs normally in sleep. (Drawing by Eleanor Sweezey.)

atism that differs little in character from that of *petit mal.**

These attacks of epileptic automatism show clearly

* The direct connections of the higher brain-stem with these two areas of cerebral cortex (prefrontal and interpretive cortex) are indicated in Figure 1. This direct connection is with the gray matter of the mind's mechanism, not with the automatic sensory-motor mechanism, as will be pointed out. Anatomical verification of this important direct relationship is to be found in the recent studies of Walle Nauta.[12]

the automatic, complex performance of which man's computer is capable. In an attack of automatism the patient becomes suddenly unconscious, but, since other mechanisms in the brain continue to function, he changes into an automaton. He may wander about, confused and aimless. Or he may continue to carry out whatever purpose his mind was in the act of handing on to his automatic sensory-motor mechanism when the highest brain-mechanism went out of action. Or he follows a stereotyped, habitual pattern of behavior. In every case, however, the automaton can make few, if any, decisions for which there has been no precedent. He makes no record of a stream of consciousness. Thus, he will have complete amnesia for the period of epileptic discharge and during the period of cellular exhaustion that follows.

Patients are quite unable to predict when these absences of the mind will come. I shall cite a few examples. One patient, whom I shall call A., was a serious student of the piano and subject to automatisms of the type called *petit mal.* He was apt to make a slight interruption in his practicing, which his mother recognized as the beginning of an "absence." Then he would continue to play for a time with considerable dexterity. Patient B. was subject to epileptic automatism that began with discharge in the temporal lobe. Sometimes the attack came on him while walking home from work. He would continue to walk and to thread his way through busy streets on his way home. He might realize later that he had had an attack because there was a blank in his memory for a part of the journey, as from Avenue X to Street Y. If Patient C. was driving a car, he would continue to drive, although he might discover later that he had driven through one or more red lights.

In general, if new decisions are to be made, the automaton cannot make them. In such a circumstance, he may become completely unreasonable and uncontrollable and even dangerous.

The behavior of these temporary automatons throws a brilliant light then, on a second mechanism, clearly distinguishable from the one that serves the mind. It is the *automatic sensory-motor mechanism*. It, too, has centrally placed gray matter in the higher brain-stem where it must have a close functional interrelationship with the mechanism for the mind. The sensory-motor mechanism has its primary localization in the higher brain-stem (see Figure 10), but the mechanism has, quite obviously, a direct relationship to the sensory and motor portions of the cerebral cortex in both hemispheres. Thus, there are two brain mechanisms that have strategically placed gray matter in the diencephalon or brain-stem, viz.: (a) the *mind's mechanism* (or highest brain-mechanism); and (b) the *computer* (or automatic sensory-motor mechanism).

When an epileptic discharge occurs in the cerebral cortex in any of the sensory or motor areas, and if it spreads by bombardment to the higher brain-stem, the result is invariably a major convulsive attack, *never*, in our experience, an attack of automatism. On the other hand, as mentioned above, a local discharge in prefrontal or temporal cortex may develop into automatism.* This is a matter of considerable functional significance, and one that has been largely overlooked. We were first aware of the differences in the manner of spread of epileptic discharge from cerebral cortex to diencephalon when

* Herbert Jasper and I showed that local epileptic discharge in the prefrontal cortex, occurring either spontaneously or when we had set it off by the electrode, might spread to the diencephalon and cause an attack of automatism that was very like the automatism of *petit mal*.[25]

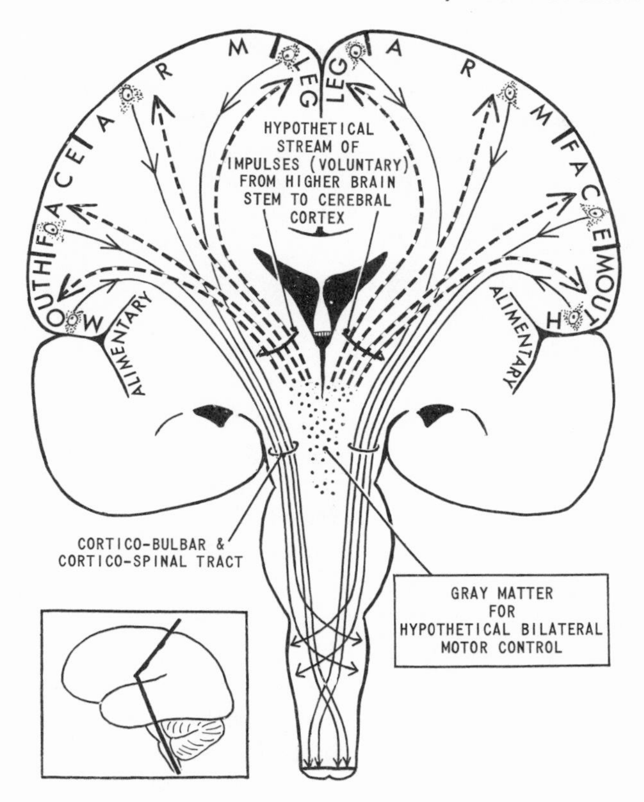

FIGURE 10. *The Automatic Sensory-Motor Mechanism.*

This much-simplified diagram outlines only the direction of the executive, or motor, messages of the mechanism that takes bilateral control of the body either under direction of the mind or automatically. It makes use of the motor cell-stations in the precentral gyrus of both sides, as shown here from leg down to face and mouth. The entire mechanism is a portion of the centrencephalic integration and coordination that makes effective mind-action possible. One may call it "man's computer." It makes available the many skills (including that of speech) that have been learned and recorded in the individual's past. It controls the behavior of the "human automaton" while the mind is otherwise occupied, or when the highest brain-mechanism is selectively inactivated, as in epileptic automatism. On the other hand, epileptic discharge within its central gray matter produces interference with its function, and calls forth active responses from the motor centers in the cortex of both hemispheres, thus producing a generalized convulsion (*grand mal*). (Drawing by Eleanor Sweezey.)

Kristian Kristiansen made his study in 1951. After examining ninety-five cases in our clinic, he pointed out that there were twenty-nine examples of seizures that began with local epileptic discharge in a motor convolution, fifty-five somatic sensory, and eleven visual sensory seizures. None developed automatism during the evolution of their attacks. Many, however, went occasionally from localized sensory or motor manifestations directly to generalized seizures.[26]

William Feindel showed that automatism is frequent (78 percent) among patients who are subject to epileptic discharges in the temporal lobe.[2] He and I showed that automatism could be produced by stimulation if the electrode was passed into the temporal lobe and on into or near the amygdaloid nucleus (see Figure 8). But this occurred only when one continued to stimulate until local epileptic discharge was produced. We assumed that this caused interference in the hippocampus on both sides and perhaps neuronal bombardment of gray matter in the higher brain-stem that went on to epileptic discharge.

Thus, from a practical point of view, a clinician may find it useful to remember that local epileptic discharge in motor or sensory gray matter areas of the cortex may spread by bombardment and so cause epileptic discharge in gray matter of the automatic sensory-motor mechanism in the higher brain-stem. This produces a major convulsion because of its activation of all the motor areas of the cortex. The cortex of one side, being pitted against the other, causes the patient to stiffen the body and limbs rather than to turn. The sensory-motor mechanism exerts activating control from its gray matter in the higher brain-stem. This acts upon the secondary gray matter in the cerebral cortex of each hemisphere, and on the tertiary gray matter in the lower brain-stem and the spinal cord. The major functional outflow of axone conducted energy is carried to the muscles in one efferent stream. During any generalized *grand mal* seizure only the automatic control of breathing, which is located in the lower brain-stem, escapes and continues its function.

When a local discharge occurs in prefrontal or temporal areas of the cortex, it may spread directly to the highest brain-mechanism by bombardment (the *mind's mechanism*). When it does this, it produces automatism. On the other hand, the sensory and motor convolutions of the cortex, when overcharged electrically, bombard the automatic sensory-motor mechanism (the *computer's mechanism*) in the higher brain-stem. One may surmise then that there must be a mind's mechanism that has direct access to prefrontal and temporal cortex but has only indirect access to sensory and motor mechanisms of the cerebral cortex.

Thus, when bombardment from the motor or sensory convolutions of the cortex becomes excessive, it may produce secondary discharge in the computer and thus a major convulsive seizure, but not an attack of automatism. When bombardment from the prefrontal or temporal areas of the cortex becomes excessive, it may produce interference in the highest brain-mechanism (the mind's mechanism) and thus cause an attack of automatism. Or it may produce a major convulsive seizure because of an additional direct connection with the computer.

No functional conclusions should be drawn from these facts except perhaps to suggest that

(1) the highest brain mechanism has direct connection with the newer temporal and prefrontal areas of cerebral cortex, and

(2) its functional connection with the older motor and sensory areas of the cortex must be an indirect one, perhaps interrupted in the computer.

11 · Centrencephalic Integration and Coordination

These two units: (1) the mechanism, the action of which is essential to the existence of consciousness; and (2) the mechanism of sensory-motor coordination, may be said to constitute the central integrating system. In their combined action they make sensory input available and motor output purposeful. They constitute a centrencephalic integrating system that unites functionally the diencephalon (higher brain-stem) with the cortex of both hemispheres. To this integrating system, sensory impressions come, and in its action, thought, and behavior find expression.

But what a difference there is between the outward evidence of epileptic discharge within the mind-mechanism and that in the automatic mechanism! Discharge in the gray matter of the mind-mechanism, like discharge in the gray matter of the speech area, results, as I have said, only in silence. There is complete interference with its function. Yet when discharge does occur in the central gray matter of the automatic sensory-motor mechanism, there is a positive response—a sudden activation of the outlying motor stations that the automaton would normally control. The result is frightening: the subject stiffens, falls, shakes, cries out, salivating and soiling himself. There is general muscular contraction. Consciousness vanishes. Only breathing, which is normally controlled by the reflex action or circuits located in the lower brain-stem, is spared. It was because of these paroxysms that the early Greeks described *epilepsia* as the curse of the gods, the divine disease.

Today, epilepsy continues to come as a curse to millions of men and women. Happily, modern treatment with medicines lessens the curse, protecting many and bringing them back to normal, productive living. And Epilepsy, though she wears the frightening mask of tragedy in her approach to each patient, takes off the mask at times before the physician who has the wit to stop and ponder her riddles. It is within her power to guide the thinking of Hippocratic disciples as she guided the master's so long ago.

The behavior of the automaton during an attack of epileptic automatism reveals what the brain without the mind and without the mind-mechanism can still do. It reveals what the moment-to-moment function of the normally active mind must be. If, as I have said, an attack of automatism falls upon a patient while he is in the act of planning a project, the automaton (which he becomes) may discharge that purpose in remarkable detail.

What has been said about epileptic automatism throws much light on what must be happening in the normal routine of our lives. By taking thought, the mind considers the future and gives short-term direction to the sensory-motor automatic mechanism. But the mind, I surmise, can give direction only through the mind's brain-mechanism. It is all very much like programming a private computer. The program comes to an electrical computer from without. The same is true of each biological computer. Purpose comes to it from outside its own mechanism. This suggests that the mind must have a supply of energy available to it for independent action.

Short-term programming of the automatic mechanism seems to serve a useful purpose in ordinary life. When I get into my car in the morning with a plan of going somewhere other than to the Montreal Neurological Institute, I have learned that I must decide in advance the streets to be followed. Otherwise, while I am thinking of something else, the automaton delivers me to the Institute.

We may assume then, that if the mind can give directions minutes in advance, it must also give directions split-seconds in advance. I assume that the mind directs, and the mind-mechanism executes. It carries the message. As Hippocrates expressed it so long ago, "the brain is messenger" to consciousness. Or, as one might express it now, the brain's highest mechanism is "messenger" between the mind and the other mechanisms of the brain.

Consider, if you will, the various functional mechanisms that operate within the brain. There is one mechanism that wakens the mind and serves it each time it comes into action after sleep. Whether one adopts a dualist or a monist hypothesis, the mechanism is essential to consciousness. It comes between the mind and the final integration that takes place automatically in the sensory-motor mechanism and it plays an essential role in that mechanism. This recalls the thinking of Hughlings Jackson in regard to the word "highest" as applied to levels of function within the brain.[3,4] That this *highest mechanism*, most closely related to the mind, is truly a functional unit is proven by the fact that epileptic discharge in gray matter that forms a part of its circuits interferes with its action selectively. During epileptic interference with the function of this gray matter, localized tentatively in Figure 9, consciousness vanishes and, with it, goes the direction and planning of behavior. That is to say, the mind goes out of action and comes into action with the normal functioning of this mechanism.

The human automaton, which replaces the man when the highest brain-mechanism is inactivated, is a thing without the capacity to make completely new decisions. It is a thing without the capacity to form new memory records and a thing without that indefinable attribute, a sense of humor. The automaton is incapable of thrilling to the beauty of a sunset or of experiencing contentment, happiness, love, compassion. These, like all awarenesses, are functions of the mind. The automaton is a thing that makes use of the reflexes and the skills, inborn and acquired, that are housed in the computer. At times it may have a plan that will serve it in place of a purpose for a few minutes. This automatic coordinator that is ever active within each of us, seems to be the most amazing of all biological computers.

By listening to patients as they describe an experiential

flashback, one can understand the complexity and efficiency of the reflex coordinating and integrative action of the brain. In it, the automatic computer and the highest brain-mechanism play interactive roles, selectively inhibitory and purposeful.

Does this explain the action of the mind? Can reflex action in the end, account for it? After years of studying the emerging mechanisms within the human brain, my own answer is "no." Mind comes into action and goes out of action with the highest brain-mechanism, it is true. But the mind has energy. The form of that energy is different from that of neuronal potentials that travel the axone pathways. There I must leave it.

13 • The Stream of Consciousness

The material used by William James in his reasoning was psychological or philosophical, rather than neurophysiological. The "stream of consciousness," he said, "is a river, forever flowing through a man's conscious waking hours."* This metaphor may be confusing. A river of water cannot be altered by the man on the bank. But thought and reason and curiosity do cause the stream of consciousness to alter its course and even change its content completely. The biological stream that is hidden away in each of us follows the command of the observer on the bank. A stream it is, and it flows inexorably onward toward the hazy sea that waits for us all at the end of life. But there the similarity between a river and the stream of consciousness breaks down.

The contents of the stream, as described in Chapter 6, are recorded in the brain, including everything to which the man on the bank paid attention, but none of the things that he ignored. His thoughts are recorded with the sensory material that he accepted. His fears are there, and his interpretations are there as well—all recorded by this extraordinary mechanism within the brain.

It is the mind (not the brain) that watches and at the same time directs. Has the mind, then, a memory of its own? No. There is no evidence to suggest it. If it has, there exists a memory mechanism of an entirely different

* As an undergraduate, majoring in philosophy at Princeton, I was much impressed by my reading of William James's *The Principles of Psychology*.[5] That was, I suppose, the beginning of my curiosity about the brain and the mind of man.

and unsuspected order. The mind has no practical need for other memory since, through the highest brain-mechanism, it can open the files of remembrance in a flash.*

* If the mind had any separate awareness while the highest brain-mechanism is inactive, it could make some use of a memory mechanism of its own. But the engram for mind-memory would have to be redefined. Instead of being the "lasting trace" left in an organism by "psychic experience," it would be the lasting trace left on psychic structure by neuronal action!

The ghost of Hamlet's father would need such a memory to converse with his own son. But certainly, no neurophysiologist is in a position to rationalize that dramatic interview born in William Shakespeare's amazing brain, and fathered by his brilliant mind.

14 · Introspection by Patient and Surgeon

In my approach to this argument I have made no reference to introspection. Instead, I have depended on neurophysiological evidence and have made special use of the evidence available after a long experience with epileptic seizures. Trying to be objective, a scientist should not trust his own introspection too far. But it might give the reader another point of view, to listen to the introspective thinking of a highly intelligent patient as it came to me while I was "tinkering" with one of his brain mechanisms. I have published the case of the patient C.H. elsewhere,* but I shall refer to it again now.

On the day of operation, I had exposed a large part of the left side of the brain under local anesthesia and had decided what should be done in the hope of freeing him from his epileptic attacks. But the focus of brain-irritation in the temporal lobe that seemed to be producing his fits was alarmingly close to where the major speech area should be, unless speech was located in the other hemisphere. So, in order to avoid the danger of producing permanent aphasia, I undertook to map out the exact position of his speech area. We have found that a gentle electrical current interferes with the function of the speech mechanism. One touches the cortex with a stimulating electrode and, since the brain is not sensitive, the patient does not realize that this has made him aphasic until he tries to speak, or to understand speech, and is unable to do so.

* *Modern Perspectives in World Psychiatry*, John G. Howells, ed. Vol. 2, p. 340. Oliver & Boyd, Edinburgh, 1968.

One of my associates began to show the patient a series of pictures on the other side of the sterile screen. C.H. named each picture accurately at first. Then, before the picture of a butterfly was shown to him, I applied the electrode where I supposed the speech cortex to be. He remained silent for a time. Then he snapped his fingers as though in exasperation. I withdrew the electrode and he spoke at once:

"Now I can talk," he said. "Butterfly. I couldn't *get* that word 'butterfly,' so I tried to *get* the word 'moth!' "

It is clear that while the speech mechanism was temporarily blocked he could perceive the meaning of the picture of a butterfly. He made a conscious effort to "get" the corresponding word. Then, not understanding why he could not do so, he turned back for a second time to the interpretive mechanism, which was well away from the interfering effect of the electric current, and found a second concept that he considered the closest thing to a butterfly. He must then have presented that to the speech mechanism, only to draw another blank.

The patient's simple statement startled me. He was calling on two brain-mechanisms alternately and at will. He had focused his attention on the cards and set himself the purpose of recognizing and naming each picture as it came along. At first each picture was inspected in the stream of consciousness. It was identified, named, and recorded. He was using areas of cerebral cortex that, at birth, had been uncommitted as to function. Evidently, the highest brain-mechanism, impelled by mind-decision, can carry out these transactions, calling upon previously established, conditioned reflexes one by one. When I paralyzed his speech mechanism, he was puzzled. Then he decided what to do. He reconsidered the concept "butterfly" and summoned the nearest thing to butterfly that was stored away in his concept mechanism. When the concept "moth" was selected and presented in the stream of consciousness, the mind approved and the highest mechanism flashed this non-verbal concept of moth to the speech mechanism. But the word for "moth" did not present itself in the stream of consciousness as he expected. He remained silent, then expressed his exasperation by snapping the fingers and thumb of his right hand. That he could do without making use of the special speech mechanism. Finally, when I removed my interfering electrode from the cortex, he explained the whole experience with a feeling of relief, using words that were appropriate to his thought. *He* got the words from the speech mechanism when *he* presented concepts to it. For the word "he," in this introspection, one may substitute the word *mind*. Its action is not automatic.

As I visualize it, a reasonable, explanatory hypothesis can be constructed as follows: because I had asked the patient to do so, he turned his attention to the naming of cards, programming the brain to that end through the highest brain-mechanism. I can say only that the decision came from his mind. Neuronal action began in the highest brain-mechanism. Here is the meeting of mind and brain. The psychico-physical frontier is here. The frontier is being crossed from mind to brain. The frontier is also being crossed from brain to mind since the mind is conscious of the meaning of the neuronal succession that determines the content of the stream of consciousness. The neuronal action is automatic as it is in any computer.

In conformity with the mind's decision, the highest mechanism sends neuronal messages to the other mechanisms in the brain. The messages go, I suppose, in the form of neuronal potentials arranged in a meaningful pattern and they are sent, in each case, to the appropriate target-gray matter. They cause the individual to turn his gaze and focus his eyes on the matter in question. They cause him to interpret what he sees, to select words that will express a meaning.

These are things that the mechanisms can do. At the same time, some brain-mechanism is admitting to consciousness the information that is relevant to the matter. It is also inhibiting all other streams of unrelated information that are flowing into the centrencephalic integrating system, so that unrelated data do not flash up in the stream of consciousness. Thus, one can conclude what must have been eliminated and therefore excluded from the patient's attention in that time and place.

This is hypothetical thinking, of course. It is clear that much is accomplished by automatic and reflex mechanisms. But what the mind does is different. It is not to be accounted for by any neuronal mechanism that I can discover.

15 · Doubling of Awareness

Consider the point of view of the patient when the surgeon's electrode, placed on the interpretive cortex, summons the replay of past experience. The stream of consciousness is suddenly doubled for him. He is aware of what is going on in the operating room as well as the "flashback" from the past. He can discuss with the surgeon the meaning of both streams.

The patient's mind, which is considering the situation in such an aloof and critical manner, can only be something quite apart from neuronal reflex action. It is noteworthy that two streams of consciousness are flowing, the one driven by input from environment, the other by an electrode delivering sixty pulses per second to the cortex. The fact that there should be no confusion in the conscious state suggests that, although the content of consciousness depends in large measure on neuronal activity, awareness itself does not.

A young South African patient lying on the operating table exclaimed, when he realized what was happening, that it was astonishing to realize that he was laughing with his cousins on a farm in South Africa, while he was also fully conscious of being in the operating room in Montreal. This illustrates what I mean. The mind of the patient was as independent of the reflex action as was the mind of the surgeon who listened and strove to understand. Thus, my argument favors independence of mind-action.

If, to the contrary, the truth is that the highest brain-mechanism is busy creating the mind by its own action, one might expect mental confusion when the neuronal record is activated by an electrode so that a stream of past consciousness is presented to the mind along with the presentation of the contemporary stream of consciousness.

One may ask this question: does the highest brain-mechanism provide the mind with its energy, an energy in such a changed form that it no longer needs to be conducted along neuraxones? To ask such a question is, I fear, to run the risk of hollow laughter from the physicists. But, nonetheless, this is my question, and the suggestion I feel myself compelled to make.

Now let me climb down from the dizzy height of the hypothetical scaffolding that has been built about the brain. Consider with me the beginning of life, leaving aside, for the moment, the question of the essential nature of the mind.

A baby brings with him into the world an active nervous system. He (or she) is already endowed with inborn reflexes that cause him to gasp and to cry aloud, and presently to search for the nipple, and to suck and swallow, and so set off a complicated succession of events within the body that will serve the purpose of nourishing it.

In the very first month you can see him—if you will take time to observe this wonder of wonders—stubbornly turning his attention to what interests him, ignoring everything else, even the desire for food or the discomfort of a wet diaper. It is evident that, already, he has a mind capable of focusing attention and evidently capable of curiosity and interest. Within a few months he recognizes concepts such as those of a flower, a dog, and a butterfly; he hears his mother speaking words, and shortly he is busy programming a large area of the uncommitted temporal cortex to serve the purposes of speech. If he is going to be right-handed, he is probably setting up the speech mechanism in the left hemisphere. He is programming, too, much of the rest of man's newly acquired temporal cortex as interpretive cortex for the recognition of *non-verbal* concepts.*

* This programming must, I suppose, begin as early as perceptions begin to be classified and as soon as attention is focused on

In doing all this, a normal, happy youngster is evidently driven by an excited curiosity to pay attention and to explore. Whatever enters the spotlight of his attention, he stores in the brain. Some of it he will be able to "remember." More of it is available when called upon by automatic mechanisms. But I dare say he stores nothing that did not first come within his focus of attention.

He makes progressive additions to, or changes in, the various concepts he is forming, choosing from what he sees and hears. The first dog to which he paid attention may have been yellow and short-haired, the next black and hairy. He modifies his concept of a dog each time he pays attention to one.

The beginning of speech is important. The first time he hears the word and imitates it, the sound will be far from his eventual pronunciation of "dog." A parrot can imitate, too. But it is not long before the infant takes another step. A dog appears in the stream of consciousness, whereupon the highest brain-mechanism carries a patterned neuronal message to the non-verbal concept-mechanism. The past record is scanned and a similar appearance recalled through the hippocampal system. The mind compares the two images that have thus appeared in the stream of consciousness, and sees similarity. There is a sense of familiarity or recognition. While all this is still

---

words. That this goes on progressively during the early years of childhood is proven by the fact that, should a completely destructive lesion come, by accident, to the major speech area in the temporal lobe, the aphasia would last no longer than a year. At the end of that period the child would begin to speak again, and before long he would speak normally, having set up a new speech area in the opposite temporal lobe. If the speech area should be destroyed after twelve years of age, the aphasia might well be permanent.

in the stream of consciousness, another patterned neuronal message is formed, made up of the remembered concept modified by the present experience. This message is sent to the speech mechanism and the word "dog" flashes up into consciousness.

Then he acts. A message is sent to the gray matter in the articulation area of the motor cortex. He speaks the word "dog" aloud—and laughs, perhaps, in conscious triumph.* I imagine that the parts in this sequence that have not become automatic are carried out by the highest brain-mechanism under the direction of the mind. I want to point out only that every learned-reaction that becomes automatic was first carried out within the light of conscious attention and in accordance with the understanding of the mind.

* However blundering my description of the neuronal and the mental action may be, it was really a magnificent performance! No puppy or dog will ever be capable of this, and certainly no parrot. Perhaps that is because each is endowed with so little uncommitted cortex. Puppy and infant may have been friends. Up to the age of six months in the learning game, the puppy had seemed to be ahead, but from now on, it was to be otherwise.

17 • What the Automatic Mechanism Can Do

Inasmuch as the brain is a place for newly acquired automatic mechanisms, it is a computer. To be useful, any computer must be programmed and operated by an external agent. Suppose an individual decides to turn his attention to a certain matter. This decision, I suppose, is an act on the part of the mind. The brain response must be somewhat as follows: The highest brain-mechanism takes immediate executive action, which causes the sensory-motor mechanism to block, by inhibition, the inflow of information that is *unrelated* to the subject of the mind's new interest. At the same time, it allows related data to pass through into the stream of consciousness. The data that enter the stream of consciousness may be sequences of sight and sound from the neighborhood, accepted with immediate interpretations, such as awareness of familiarity or of danger. Relevant memories may be added automatically from the individual's past experience.*

The automatic sensory-motor mechanism itself comes to be conditioned, as the years pass, for many purposes.

* I suggest that this selection of what enters the stream of consciousness is a function of the automatic sensory-motor mechanism because, during an attack of epileptic automatism, when the automaton is in full control and the highest brain-mechanism is paralyzed, this selection still continues, at least in regard to the purpose for which the mechanism has been most recently programmed. Thus, the automaton can walk through traffic as though he were aware of all that he hears and sees, and so continue on his way home. But he is aware of nothing and so makes no memory record. If a policeman were to accost him he might consider the poor fellow to be walking in his sleep.

It coordinates the action of many semi-separable mechanisms within the brain, such as reading, writing, speaking, and the dextrous skills. As time passes, it learns to take over more and more of the body's behavior. Each skill, acquired in the light of conscious attention, soon becomes automatic and runs itself even more skillfully than the individual could carry it out by conscious direction.

If decisions as to the target of conscious attention are made by the mind, then the mind it is that directs the programming of all the mechanisms within the brain. A man's mind, one might say, is the person. He walks about the world, depending always upon his private computer, which he programs continuously to suit his ever-changing purposes and interest.

Mind, brain, and body make the man, and the man is capable of so much! He is capable of comprehension of the universe, dedication to the good of others, planned research, happiness, despair, and eventually, perhaps, even an understanding of himself. He can hardly be subdivided. Certainly, mind and brain carry on their functions normally as a unit.

The neurophysiologist's initial undertaking should be to try to explain the behavior of this being on the basis of neuronal mechanisms alone—this biped who falls on his knees to pray to his god and rises to his feet to lead an army, or to write a poem, or to dig a ditch, or to thrill to the beauty of the sunrise, or to laugh at the absurdities of this world. To me, it seems more and more reasonable to suggest, as I did at the close of the Thayer Lectures at Johns Hopkins in 1950 (unpublished), that the mind may be a distinct and *different essence.* A science reporter, who was present at the lectures, used those words in his report as though that were the conclusion. I was not ready for that then. I am hesitant still. But, look with me for a moment at our world.

If one is to make a judgment on the basis of behavior, it is apparent that man is not alone in the possession of a mind. The ant (whose nervous system is a highly complicated structure), as well as such mammals as the beaver or the dog or chimpanzee, shows evidence of consciousness and of individual purpose. The brain, we may assume, makes consciousness possible in him too. In all of these forms, as in man, memory is a function of the brain. Animals, particularly, show evidence of what may be called racial memory. Secondly, new memories are acquired in the form of conditioned reflexes. In the case of man, these preserve the skills, the memory of words, and the memory of non-verbal concepts. Then there is, in man at least, the third important form of memory, experiential memory, and the possibility of recalling the stream of consciousness with varying degrees of completeness. In this form of memory and in speech, the convolutions, which have appeared in the temporal lobe of man as a late evolutionary addition, are employed as "speech-cortex and interpretive-cortex."

Perhaps I should recapitulate. I realized as early as 1938 that to begin to understand the basis of consciousness one would have to wait for a clearer understanding of the neuronal mechanisms in the higher brain-stem. They were obviously responsible for the neuronal integrative action of the brain that is associated with consciousness. Since then, electrical stimulation and a study of epileptic patterns in general have helped us to distinguish three integrative mechanisms. Each has a major area or nucleus of gray matter (within the higher brain-stem), an aggregation of nerve cells that may be activated or paralyzed.

(a) Highest Brain-Mechanism

The function of this gray matter (Figure 9) is to carry out the neuronal action that corresponds with action of the mind. Proof that this is a mechanism in itself depends on the facts that injury of a circumscribed area in the higher brain-stem produces invariable loss of consciousness, and that the selective epileptic discharge, which interferes with function within this gray matter, can produce the unconsciousness (seen in epileptic automatism) without paralysis of the automatic sensory-motor control mechanism located nearby.

(b) Automatic Sensory-Motor Mechanism

The function of this gray matter (Figure 10) is to coordinate sensory-motor activity previously programmed by the mind. This biological computer mechanism carries on automatically when the highest brain-mechanism is selectively inactivated. It causes a major convulsive attack, *grand mal*, by activation of cortical motor convolutions, when epileptic discharge occurs in its central gray matter.

(c) The Record of Experience

The function of the central gray matter of this mechanism (Figure 11), as shown by electrode activation, is to recall to a conscious individual the stream of consciousness from past time.

As I have said, the function of each of these three areas of gray matter can now be recognized. Each plays a role in the normal centrencephalic integration and coordination, which is prerequisite to, and which accompanies, consciousness. As has been pointed out, this integrative activity is arrested by any form of general functional interference with neuronal activity in the region outlined in Figure 9. The central gray matter for the automatic mechanism (b) and that of the mind's mechanism (a) are shown in separate figures (9 and 10), since the two are capable of separate function and I cannot picture the exact anatomical relationship.

Thanks to patients who have told us about the stream of consciousness presented to them while the surgeon's

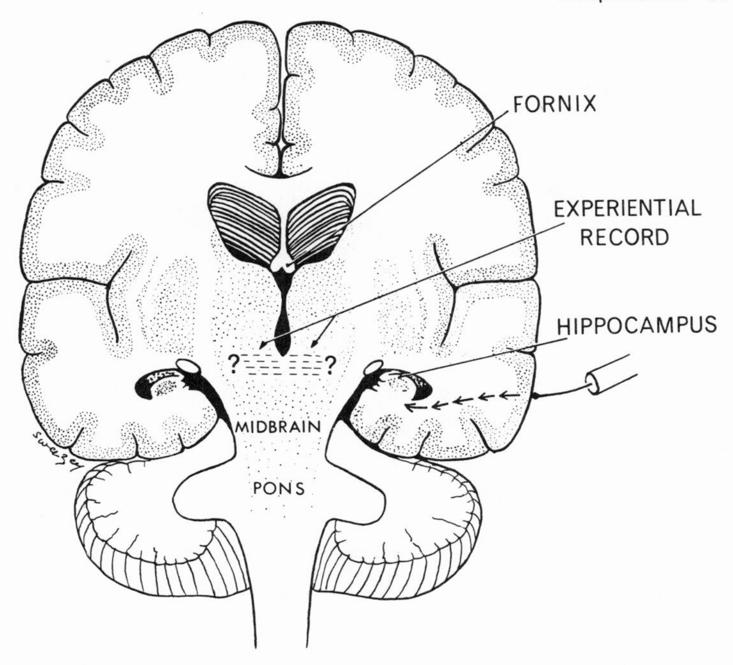

FIGURE 11. *The Experiential Record of Consciousness and Its Reactivation.*

A stimulating electrode is shown here applied to a convolution in the interpretive-cortex. It recalls the past by activating a functional circuit (experiential response). The circuit has yet to be completely defined, but may well include the hippocampus and fornix.* (Drawing by Eleanor Sweezey.)

* The beginning of the pathway of neuronal activation is shown by the line of arrows from beneath the electrode. It may be that it passes directly through white matter into the higher brain-stem. Or it may enter the hippocampus and pass along the fornix, which conducts from hippocampus to brain-stem. The hippocampus evidently plays an essential role, together with the interpretive-cortex, in the function of scanning the experiential record and recalling memories. Removal of one hippocampus affects memory little. Removal of both abolishes the man's ability

activating electrode was still in place, we can conclude with certainty what data the above integrative mechanisms inhibit (and, thus, eliminate) and what they present to the mind, and, thus, cause the brain to record. The imprint of memory's engram is somehow added during neuronal action. Conscious attention seems to give to that passage of neuronal impulses permanent facilitation for subsequent passage of potentials along the neuronal connections in the same pattern. Thus, a recall engram is established. This, one may suggest, is the real secret of learning. It is effective in the establishment of experiential memory, as well as of the memory based on conditioned reflexes.

If this is so, then it was correct to assume that the record of the stream of consciousness is not in some duplicated separate memory apparatus such as that of the hippocampus apparatuses, but in the operational mechanism of the highest brain-mechanism. The hippocampal gyrus in each hemisphere is part of the mechanism of scanning and recall. Therefore, there must be a laying down in each hippocampus of some sort of duplicated key-of-access to the record of the stream of consciousness.[27]

---

to recall past experience, either voluntarily or for the purposes of automatic interpretation. In the case of a simple bilateral loss of the hippocampi, other forms of memory are retained, e.g., memory of language, and of skills, and of non-verbal concepts. (See Penfield and Mathieson,[27] 1974.)

19 • Relationship of Mind to Brain—A Case Example

Consciousness can be present whenever the highest brain-mechanism is normally active, even though adjacent parts of the brain are inactivated by some abnormal influence. A patient, whom I was called upon to see in Moscow under dramatic circumstances in 1962, illustrated the fact that conscious understanding may be present when motor control has been lost completely, or almost completely, and when the brain is not capable of making a permanent record of the stream of consciousness.

The patient was the brilliant physicist, Lev Landau. Only intensive nursing had kept him alive during six weeks of complete unconsciousness following a head injury in an automobile crash. On my first examination of the patient, I agreed that he was completely unconscious. I then recommended a minor diagnostic operation (ventriculography). His limbs were paralyzed; his eyes were open but apparently unseeing. Next morning, when I entered his room to examine him again, I was accompanied by his wife. She preceded me and, sitting down at the bedside, she talked to him, telling him that I had suggested to the Soviet surgeons that he should have a brain operation. As I stood silent, watching over her head, I became aware of a startling change in the patient. He lay unmoving still, as on the previous night. But his eyes, which had been deviated from each other then, were focused now in a normal manner. He seemed to be looking at her. He appeared to hear, and see, and to understand speech! How could this be? She came to the end of her explanation and was silent. His eyes then moved upward to focus quite normally on me. I moved my head from side to side. The eyes followed me. No doubt about it! Then they swung apart again and he appeared, as he had the night before, to be unconscious.

It was clear that the man had returned to consciousness. He had been able to hear, see, understand speech, but not to speak. He could not move, except to focus and turn his eyes briefly. Perhaps I should explain that he and his wife had been separated for a time. It was our talk of possible operation that had led to her being summoned to Moscow and to the hospital. She was seeing him that morning for the first time since the accident.

It was thrilling to realize what had happened. He had been roused by her presence and probably understood her message. Evidently the hemispheres above the higher brain-stem, with their speech and visual and auditory mechanisms, had not been injured. The to-and-fro exchange between brain-stem and cortex was free, but when he sent neuronal messages out to the peripheral motor nuclei in the lower brain-stem and spinal cord, none could pass the block at the level of the hemorrhage in the midbrain. None, that is, except those to the eye movement center, which is highest of all the peripheral centers for motor control. If he was conscious, he must have sent down many other messages that would normally have flashed outward to the muscles. His wife was an appealingly handsome woman. His mind may well have sent a message intended to cause his hand to take hers. But his hand lay motionless.

However that may be, I went back to the other doctors and we decided that no operation was necessary. I had seen the first sure sign of recovery. He was transferred at once from the outlying hospital, in which he had been nursed so magnificently, to the Moscow Neurosurgical

Institute, where he would have the great advantage of supervision by Professor B. G. Egorov. Physiotherapy was begun at once, and I learned later that there was slow but continuous and progressive recovery from that day onward.*

For the first six weeks after the accident, Landau's fellow physicists, most of them his disciples, had joined the nurses and doctors in their gallant effort to keep the patient breathing and capable of recovery if that should prove possible. This man, who had already been awarded the Lenin Prize for his contributions to physics, was given the Nobel Prize during his convalescence. He and his wife were happy together and she was with him on the special occasion of his acceptance of the award.

I saw him nine months later, on my return from a visit to the university hospitals of China. He had been transferred to a convalescent hospital. The neurologist, Professor Propper Graschenkov, took me, in company with Landau's collaborator, the brilliant Professor Lifshitz, who had shared the Lenin Prize with his master, to visit him. I quote now from my journal note, made a few days later: "Landau was sitting up in bed in a fresh white shirt looking at me anxiously. He was a handsome person with a fine head. He had an air of understanding. He was stubborn in his confusion and I could appreciate why, several months earlier, he had been depressed and had even asked Lifshitz to bring him poison." Landau spoke excellent English and did so eagerly. "When I asked him if he had received the Lenin Prize, he said 'yes,' but he could not tell me when. When I asked him if someone had shared it with him he looked around and, finding

* An account of this case was published by a science reporter in an interesting book for lay readers: *The Man They Would Not Let Die*, by Alexander Dorozinski, Macmillan, New York, 1965.

Lifshitz standing with the others, smiled and pointed to him—'I think he and I shared it.' " The relationship of the two men was somewhat like that of David and Jonathan. "Lifshitz feels he has lost his closest friend and his leader."

The final diagnosis was as follows: A head injury that resulted in a small hemorrhage in the conducting tracts (white matter) of the higher brain-stem. During the six weeks following his accident, the hemorrhage had gradually absorbed through the brain's own circulation. Eventually, neuronal communication had been restored through the nerve fibers of the brain-stem. There remained only a difficulty in recalling the recent past. His confusion was due to that.

It is, of course, only a matter of conjecture exactly when consciousness did return. Before he could show that he was conscious, it was necessary for him to regain control of some portion of his motor system. Evidently, control of the movement of his eyes returned before anything else. During our first interview, while Mrs. Landau was explaining the situation to him, and when he looked at me so searchingly, I concluded that the whole explanation continued to be clear in his mind. He was remembering what she had said at the beginning as well as what she was saying at the close of her explanation. For that, the mind does not need the brain's mechanism of experiential recall. It needs only the mechanism of the stream of consciousness, which was normally active.

When I saw him at the second interview nine months later he did not remember me or anything that had happened to himself for some months after his transfer to the Moscow Neurosurgical Institute. The scanning and recall of past experience can only be carried out by the mechanism in which at least one hippocampus of one temporal lobe must play an essential role (Figure 11). And that mechanism can carry out this recall only for periods in which it had been normal and, thus, able to form its own clues to the stream of consciousness, its duplicate "keys-of-access."[27]

I learned still later that Landau did continue to improve in the year that followed, and was able to tutor his son in preparing for his university entrance examinations. Great recognition came to this man whose mathematical genius had been likened to that of Einstein. His countrymen rejoiced at his recovery but the "depression" returned to him. Perhaps he realized that his brain could no longer serve him as it had.

I have described this case in some detail because it, like many others, demonstrates how it is that when consciousness is present, the highest brain-mechanism is used to activate and employ other brain-mechanisms that are capable of normal function. Beyond that, it bears out our conclusion that the mind can hold the data that have come to it during this focusing of attention and while the mechanism of the stream of consciousness is moving forward. But the mind, by itself, cannot recall past experience unless the brain's special mechanism of scanning and recall is functioning normally. In such a case as this, in which damage brings to the brain a small area of paralyzing interference, one's knowledge of how brain mechanisms coordinate and integrate is put to the test. More than that, such accidental experiments point the way toward clearer understanding of how the business of the brain is transacted. Human physiology can only wait for guidance from such accidental experiments.

Hughlings Jackson remarked in his Hunterian Oration in 1872 that "Medical men, since they, only, witness the results of the experiments of disease on the nervous system of man, will be looked to more and more for facts bearing on the physiology of the mind." Since he used the daring phrase "physiology of the mind," he must have thought that the brain did, or would someday, explain the mind. After a century of study by neurologists who observe the paralyzed and the epileptic, and neurosurgeons who excise and stimulate the convolutions of the brain, and electrophysiologists who record and conduct experiments, surely the time has come to ask that question and other questions.

Has the brain explained the mind? If it has, does the brain do so by the simple performance of its neuronal mechanisms, or by supplying energy to the mind? Or both? Does it supply the mind with energy and at the same time provide it with basic neuronal mechanisms that are related to consciousness?

Sherrington concluded: "That our being should consist of two fundamental elements offers, I suppose, no greater inherent improbability than that it should rest on one only."

The challenge that comes to every neurophysiologist is to explain in terms of brain mechanisms all that men have come to consider the work of the mind, if he can. And this he must undertake freely, without philosophical or religious bias. If he does not succeed in his explanation, using proven facts and reasonable hypotheses, the time should come, as it has to me, to consider other possible explanations. He must consider how the evidence can be made to fit the hypothesis of two elements as well as that of one only.

Man himself, with his enormous brain, his new convolutions, and his capacity of programming his own computer, is a highly improbable end-result to anticipate. But end-result he is, in our time. The wonder is not so much that man has come out of the slow, slow process of evolution with its chance variations and survival values, but that there should be a universe at all, with its laws and plan and apparent purpose. One suspects that given time unlimited, creation begun on any kindly planet would come to much the same end-result.

The certainty, to my mind, is this: When we come to understand man, we will see that the nature of the mind and the nature of the mind's energy is simple and easily comprehensible. However, that shows only that I am an optimist.

Charles Sherrington and his pupils analyzed the integrative action of *inborn reflexes* in the nervous system of *unconscious* animals. He described the remarkable neuronal machine much as it is possessed by man in common with laboratory mammals. It provides for reflex standing, walking, and reacting to what the individual sees, hears, feels, and smells in his environment. It is an amazing machine. When Sherrington stimulated the cerebral cortex electrically and activated one or another motor mechanism he showed that there was a *facilitation effect.* Subsequent passage of a weaker electric current could produce the same reaction, although facilitation of the reaction was temporary, lasting only a matter of seconds or minutes. (See Sherrington.[32])

Ivan Pavlov and his disciples studied *conscious animals.*[11] He described conditioned reflexes that the animal acquired because anticipation caused it to pay attention to the experience. These reflexes were established and the patterns were preserved by permanent facilitation in the cerebral cortex and nuclear gray matter of the adjacent higher brain-stem. This was recognized as the physical basis of the learning of skills and simple forms of reactive behavior. It was assumed to apply to man as well as animal. Such permanent facilitation also seems to take place, I repeat, in the light of focused attention.

Since then, further facts have come to light in the study of *conscious man.* There is the record of conscious experience that we have discussed. It makes possible voluntary and automatic recall of past experiences, and it includes those things to which the individual paid attention, nothing he ignored. One can only conclude that conscious attention adds something to brain-action that would otherwise leave no record. It gives to the passage of neuronal potentials an astonishing permanence of facilitation for the later passage of current, as though a trail had been blazed through the seemingly infinite maze of neurone connections. The same principle applies to the acquisition of speech skills and the storing of non-verbal concepts. Permanent facilitation of a patterned sequence in these brain mechanisms is established only when there is a focusing of attention on the phenomenon that corresponds to it in consciousness.

If previous decision regarding the focusing of attention is made in the mind, then it is the mind that decides when the facilitating engram is to be added. One may assume that it is the highest brain-mechanism that initiates the brain action associated with that decision. One may assume, too, that the engram is simultaneously added to conditioned reflexes and to the sequential record of conscious experience.

Is there any evidence of the existence of neuronal activity within the brain that would account for what the mind does?

Before venturing to give an answer, it may be of interest to refer again to action that the mind seems to carry out independently, and then to reconsider briefly our experience with stimulation of the cortex of conscious patients and our experience of what effects are produced by epileptic discharge in various parts of the brain. This should give some clue if there is a mechanism that explains the mind.

(a) What the Mind Does

It is what we have learned to call the mind that seems to focus attention. The mind is aware of what is going on. The mind reasons and makes new decisions. It understands. It acts as though endowed with an energy of its own. It can make decisions and put them into effect by calling upon various brain mechanisms. It does this by activating neurone-mechanisms. This, it seems, could only be brought about by expenditure of energy.

(b) What the Patient Thinks

When I have caused a conscious patient to move his hand by applying an electrode to the motor cortex of one hemisphere, I have often asked him about it. Invariably his response was: "I didn't do that. You did." When I caused him to vocalize, he said: "I didn't make that sound. You pulled it out of me." When I caused the record of the stream of consciousness to run again and so presented to him the record of his past experience, he marvelled that he should be conscious of the past as well as of the present. He was astonished that it should come back to him so completely, with more detail than he could possibly recall voluntarily. He assumed at once that, somehow, the surgeon was responsible for the phenomenon, but he recognized the details as those of his own past experience. When one analyzes such a "flashback" it is evident, as I have said above, that only those things to which he paid attention were preserved in this permanently facilitated record.

(c) What the Electrode Can Do

I have been alert to the importance of studying the results of electrode stimulation of the brain of a conscious man, and have recorded the results as accurately and completely as I could. The electrode can present to the patient various crude sensations. It can cause him to turn head and eyes, or to move the limbs, or to vocalize and swallow. It may recall vivid re-experience of the past, or present to him an illusion that present experience is familiar, or that the things he sees are growing large and coming near. But he remains aloof. He passes judgment on it all. He says "things *seem* familiar," not "I have been through this before." He says, "things are growing larger," but he does not move for fear of being run over. If the electrode moves his right hand, he does not say, "I wanted to move it." He may, however, reach over with the left hand and oppose his action.

There is no place in the cerebral cortex where electrical stimulation will cause a patient to believe or to decide. Of course, there are the areas devoted to speech, whose function is arrested without the production of any distant positive response. There are also, areas of gray matter in the higher brain-stem that the surgeon's stimulating electrode does not explore. Yet epileptic discharge can take place in any area of gray matter, from diencephalon to spinal cord (except perhaps the cerebellum). The Jacksonian march of discharge, from one nucleus of the diencephalon to another, does occur as it does in the cortex.[13]

(d) Activation by Epileptic Discharge

There is *no* area of gray matter, as far as my experience goes, in which local epileptic discharge brings to pass what could be called "mind-action." In 1954 my associate, Herbert Jasper, and I reviewed our clinical experience in the field of epilepsy with this question in mind. There were only two cases in which one might have considered the beginning of an attack to be an example of "forced thinking" or "intellectual aura."*

Since this was the only suspicion of such a thing, I am forced to conclude that there is no valid evidence that either epileptic discharge or electrical stimulation can activate the mind.

If one stops to consider it, this is an arresting fact. The record of consciousness can be set in motion, complicated though it is, by the electrode or by epileptic discharge. An illusion of interpretation can be produced in the same way. But none of the actions that we attribute to the mind has been initiated by electrode stimulation or epileptic discharge. If there were a mechanism in the brain that could do what the mind does, one might expect that the mechanism would betray its presence in a convincing manner by some better evidence of epileptic or electrode activation. But, of course, I must admit that all of this is only negative evidence.

Let us consider what light our positive neurophysiological evidence can throw on the nature of man's being: If there is only one fundamental element in man's being, then neuronal action within the brain must account for all the mind does. The "indispensable substratum" of consciousness is in the higher brain-stem, as described in Chapter 5. The highest brain-mechanism's activity seems to correspond with that of the mind, as we have discussed in Chapter 12. This mechanism, as it goes out of action in sleep and resumes action on waking, may

* See pages 468-469 of our book:[25] The patient, W.S., at the beginning of an attack, found himself saying to himself that "he told somebody or other to do this or that." J.J. tried to describe what had happened to his thinking thus: "There was a slice of bread on the table. I thought it was necessary to turn or move the bread."

switch off the mind and switch it on. It may, one can suggest, do this by supplying and by taking away the energy that might come to the mind from the brain. But to expect the highest brain-mechanism or any set of reflexes, however complicated, to carry out what the mind does, and thus perform all the functions of the mind, is quite absurd.

If that is true, what other explanation can one propose? Only that there is, in fact, a second fundamental element and a second form of energy. But, on the basis of mind and brain as two semi-independent elements, one would still be forced to assume that the mind makes its impact upon the brain through the highest brain-mechanism. The mind must act upon it. The mind must also be acted upon by the highest brain-mechanism. The mind must remember by making use of the brain's recording mechanisms. The mind is present whenever the highest brain-mechanism is functioning normally.

If there are two elements, then energy must be available in two different forms. There is force that is made available through neuronal conduction in the brain. Is there force that is available to the mind, which has no such circuits? Can chemical action in nerve cells result in brain action on the one hand and in mind action on the other? Electricity was first revealed to science while it was being conducted along the nerves of living organisms. Physicists might well consider our questions seriously today, if only out of gratitude!

During brain action a neurophysiologist can surmise where the conduction of potentials is being carried out and its pattern. It is not so in the case of what we have come to call mind-action. And yet the mind seems to act independently of the brain in the same sense that a programmer acts independently of his computer, however much he may depend upon the action of that computer for certain purposes.

Now that one begins to see the outlines of the partially separable mechanisms taking shape within the brain, and can begin to understand what the reflex mechanisms are capable of doing, the time has come to propose hypotheses that would account for the transactions of the mind that are not so explained and to choose the most reasonable.

For my own part, after years of striving to explain the mind on the basis of brain-action alone, I have come to the conclusion that it is simpler (and far easier to be logical) if one adopts the hypothesis that our being does consist of two fundamental elements. If that is true, it could still be true that energy required comes to the mind during waking hours through the highest brain-mechanism.

Because it seems to me certain that it will always be quite impossible to explain the mind on the basis of neuronal action within the brain, and because it seems to me that the mind develops and matures independently throughout an individual's life as though it were a continuing element, and because a computer (which the brain is) must be programmed and operated by an agency capable of independent understanding, I am forced to choose the proposition that our being is to be explained on the basis of two fundamental elements. This, to my mind, offers the greatest likelihood of leading us to the final understanding toward which so many stalwart scientists strive.

So many questions still confront us! But to ask them is the first step toward solution. I am confident that they will be answered in time. After adopting the dualist hypothesis one can quite logically call upon the physicists for help. Can electrical energy take two forms? What is the nature of the mind? Has it a structure? Can there be energy without structure? What is electricity? Whatever the answers to these questions may be, the mind is present.

As Aristotle expressed it, the mind is "attached to the body." The mind vanishes when the highest brain-mechanism ceases to function due to injury or due to epileptic interference or anesthetic drug. More than that, the mind vanishes during deep sleep.

What happens when the mind vanishes? There are two obvious answers to that question; they arise from Sherrington's two alternatives—whether man's being is to be explained on the basis of one or two elements. If the first alternative is chosen, the mind no longer exists when it vanishes, since it is only a function of brain action. Mind is recreated each time the highest brain-mechanism goes into normal action. In this case, one must try to see the mind as the action of a specialized mechanism of the brain, the mechanism that I have called the "highest."

Or, if one chooses the second, the dualistic alternative, the mind must be viewed as a basic *element* in itself. One might, then, call it a *medium*, an *essence*, a *soma*. That is to say, it has a *continuing existence*. On this basis, one must assume that although the mind is silent when it no longer has its special connection to the brain, it exists in the silent intervals and takes over control when the highest brain-mechanism does go into action.

Thus, it would seem that this specialized brain-mechanism switches off the power that energizes the mind each time it falls asleep. It switches on the mind when it wakens. This is the daily automatic routine to which all mammals are committed and by which the brain recovers from fatigue.

The highest brain-mechanism switches on this *semi-independent element*, which instantly takes charge during wakefulness, and switches it off in sleep. Does this seem to be an improbable explanation? It is not so improbable, to my mind, as is the alternative expectation—that the highest brain-mechanism should itself understand, and reason, and direct voluntary action, and decide where attention should be turned and what the computer must learn, and record, and reveal on demand.

But in the case of either alternative, the mind has no memory of its own as far as our evidence goes. The brain, like any computer, stores what it has learned during active intervals. All of its records are instantly available to the conscious mind throughout the person's waking life, and in a distorted fashion during the dreams of the half-asleep state.

21 • Comprehensibility

A century of scientific progress has passed since Hughlings Jackson suggested that there were high levels of functional organization in the brain. He seemed to consider the highest as most closely related to the mind. Since his time, various partially independent mechanisms have been identified and mapped in the cerebral cortex and the higher brain-stem. None of them can explain the mind. The mind remains a mystery.

I have told the story here of the progress of one pilgrim as he stumbled, sometimes blindly but always hopefully, toward a clearer understanding of what seemed to be the physical basis of the mind. It is almost forty years now since he happened upon the fact that a gentle electric current applied to the interpretive cortex of the temporal lobe could summon a flashback, thus activating the stream of consciousness from the past. Gradually, over the years, he has made critical reports of the nature of these experiential responses. These and the other results of electrical stimulation are reliable data, not opinions. The effects of epileptic discharge are clues to understanding. Valid evidence has been presented that the integrative neuronal action, which makes consciousness possible, is localized in the higher brain-stem rather than in the cerebral cortex.

Now, I have suggested in this essay that there is a special form of energy that activates the mind during waking hours and that it must be derived somehow from neuronal energy. Hippocrates foretold the discovery of the highest brain-mechanism when he said, "to consciousness the brain is messenger." The highest brain-mechanism is the messenger between the mind and the other brain-mechanisms. Or, to express it another way, the highest brain-mechanism is the mind's executive. Somehow, the executive accepts direction from the mind and passes it on to various mechanisms of the brain. Thus, it passes on the short-term purposes of the mind to the automatic sensory-motor mechanism, which, in turn, carries a man through much of his apparently conscious behavior in life. These two, the highest mechanism and the sensory-motor mechanism, coordinate sensory-input and motor-output in accordance with the purpose and the direction-of-attention of the mind. They manage the employment of the various skills, including that of speech. Together they carry out the central integrative activity of the brain.

The foregoing statements are, of course, hypotheses in regard to the physical basis of the mind. They will serve to point the way ahead while one waits for an understanding of how messages are sent along neuronal circuits. That such mechanisms do exist, and such messages are carried, is proved by the evidence of epileptic patterns of discharge. It is proved also by electrical stimulation and by the many proofs of clinical experience.

Other, younger men will have to reconsider critically the localization of the gray matter that is activated in experiential responses (Figure 11), and fill in the detail for Figures 9 and 10. They must proceed with the elaboration of hypotheses to explain the "how" of neuronal action during the focusing of attention. Finally, fresh explorers must discover how it is that the movement of potentials becomes awareness, and how purpose is translated into a patterned neuronal message. Neurophysiologists will need the help of chemists and physicists in all this, no doubt.

There are many men of differing disciplines who can use of these data, whether they find it reasonable to attempt to fit them into the hypothesis that the brain explains the mind, or whether they conclude, as I have done, that the mind is a separate but related element. One of these two "improbabilities" must be chosen.

Taken either way, the nature of the mind presents the fundamental problem, perhaps the most difficult and most important of all problems. For myself, after a professional lifetime spent in trying to discover how the brain accounts for the mind, it comes as a surprise now to discover, during this final examination of the evidence, that the dualist hypothesis seems the more reasonable of the two possible explanations.

Since every man must adopt for himself, without the help of science, his way of life and his personal religion, I have long had my own private beliefs. What a thrill it is, then, to discover that the scientist, too, can legitimately believe in the existence of the spirit!

In the remainder of this chapter I shall at times speak less as a physiologist and more as a physician who, in addition to his addiction to science, is concerned about his patients, his family, and himself. But I shall do my best to take critical judgment with me each time I step "outside the boundaries of natural science."

Possibly the scientist and the physician could add something by stepping outside the laboratory and the consulting room to reconsider these strangely gifted human beings about us. Where did the mind—call it the spirit if you like—come from? Who can say? It exists. The mind is attached to the action of a certain mechanism within the brain. A mind has been thus attached in the case of every human being for many thousands of generations, and there seems to be significant evidence of heredity in the mind's character from one generation to the next and the next. But at present, one can only say simply and without explanation, "the mind is born."

Physicians, whose task it is to deal with the whole man, take a unique view of him. They have long been aware of the unexplained dichotomy (the functional split) between mind and body. Indeed, they have learned, as the saying goes, to "treat the mind as well as the body." They are well aware that body, brain, and mind make up the child. They develop together and yet they seem to remain apart as the years pass. These three, in a sort of ontogenetic symbiosis, go through life together. Each is useless without the other two. Mind takes the initiative in exploring the environment.

Mind decides what is to be learned and recorded. The child grows and the mind comes to depend more and more on the memory and the automatic patterns of action stored away in the brain's computer. The mind conditions the brain. It programs the computer so that it can carry out an increasing number of routine performances. And so, as years pass, the mind has more and more free time to explore the world of the intellect, its own and that of others.

If one were to draw curves to show the excellence of human performance, those of the body and the brain would rise, each to its zenith, in the twenties or the thirties. In the forties, the curves would level off and begin to fall, for there are pathological processes, some peculiar to the body and some to the brain, that inevitably slow them down as though with weights of lead. Thus the curves of excellent physical performance slope downward toward inevitable zero. The Psalmist saw all this 3,000 years ago when he wrote: "The days of our years are threescore years and ten" or, "by reason of strength . . . fourscore years." Man's span of life is predetermined. The legs grow weak. The memory-record, so readily available in early years, opens its "file" more slowly and reluctantly as the years pass. In the end, the brain may even fail, at times, to make any record at all of current events. Senescence is a symptom of failing performance by the body and the brain. Thus it enters one's life in various forms.

In contrast to the other two, the mind seems to have no peculiar or inevitable pathology. Late in life, it moves to its own fulfillment. As the mind arrives at clearer understanding and better balanced judgment, the other two are beginning to fail in strength and speed.

Here, as I approach the end of this study, is a further suggestion from the physician's point of view. It is an observation relevant to any inquiry into the nature of man's being, and in conformity with the proposition that the mind has a separate existence. It might even be taken as an argument for the feasibility and the possibility of immortality!

"What becomes of the mind after death?"

That question brings up the other question so often asked: "Can the mind communicate directly with other minds?" As far as any clearly proven scientific conclusion goes, the answer to the second question is "no." The mind can communicate only through its brain-mechanisms. Certainly it does so most often through the mechanism of speech. Nonetheless, since the exact nature of the mind is a mystery and the source of its energy has yet to be identified, no scientist is in a position to say that direct communication between one active mind and another cannot occur during life. He may say that unassailable evidence of it has not yet been brought forward.

Direct communication between the mind of man and the mind of God is quite another matter. The argument, in favor of this, lies in the claim, made by so many men for so long a time that they have received guidance and revelation from some power beyond themselves through the medium of prayer. I see no reason to doubt this evidence, nor any means of submitting it to scientific proof.

Indeed, no scientist, by virtue of his science, has the right to pass judgment on the faiths by which men live and die. We can only set out the data about the brain, and present the physiological hypotheses that are relevant to what the mind does.

Now we must return, however reluctantly, to the first question: When death at last blows out the candle that was life, the mind seems to vanish, as in sleep. I said "seems." What can one really conclude? What is the reasonable hypothesis in regard to this matter, considering the physiological evidence? Only this: the brain has not explained the mind fully. The mind of man seems to derive its energy, perhaps in altered form, from the highest brain-mechanism during his waking hours. In the daily routine of a man's life, communication with other minds is carried out indirectly through the mechanisms of the brain. If this is so, it is clear that, in order to survive after death, the mind *must* establish a connection with a source of energy other than that of the brain. If not, the mind must disappear forever as surely as the brain and the body die and turn to dust. If, however, during life, when brain and mind are awake, direct communication is sometimes established with the minds of other men or with the mind of God, then it is clear that energy from without can reach a man's mind. In that case, it is not unreasonable for him to hope that after death the mind may waken to another source of energy.

I mean that if the active mind of a man does communicate with other active minds, even on rare occasions, it could do so only by the transfer of some form of energy from mind to mind directly. Likewise, if the mind of man communicates with the mind of God directly, that also suggests that energy, in some form, passes from spirit to spirit. It is obvious that science can make no statement at present in regard to the question of man's existence after death, although every thoughtful man must ask that question. But, when the nature of the energy that activates the mind is discovered (as I believe it will be), the time may yet come when scientists will be able to make a valid approach to a study of the nature of a spirit other than that of man.

Man has no cause to fear the truth. It can, in the end, only fortify the valid creeds by which he lives. Deep in his nature is the urge to explore, and to learn, and to adopt a creed that will give him reassurance. The common man has a personal *credo* too. It is apt to differ, if only a little, from that of his fellows. And therein lies the strength and hope of the race.

The facts and hypotheses discussed here may well be of use in many fields of specialized thinking such as religion, philosophy, and psychiatry, as well as physics, chemistry, and medicine. Whether the mind is truly a separate element or whether, in some way not yet apparent, it is an expression of neuronal action, the decision must wait for further scientific evidence. We have discussed here only one body of that evidence. But, since it is incomplete, one must consider two hypotheses of explanation.

Meanwhile, there is a practical problem confronting man. He must learn to control his own social evolution, and I speak, for the moment, as a concerned physician rather than as a basic scientist. This is the pressing problem of human destiny. This problem will be solved only through more fundamental understanding. Comprehension will make clear the way of wisdom.

Biophysicists might well reflect that after life appeared in the long, long story of creation, evidence of self-awareness made its first appearance along with the appearance of the complicated brain. Thus, late in the process of biological evolution, *consciousness*—a new and very different phenomenon—had presented itself. This resulted in the appearance of a new world, created by the mind of man, in which there was understanding and reason and, eventually, onrushing social evolution!

What a challenge is here for man to face, a problem no less vast than that to be glimpsed in outer space! It was a physicist, Albert Einstein, who in a moment of understanding exclaimed: "The mystery of the world is its comprehensibility."

I have no doubt the day will dawn when the mystery of the mind will no longer be a mystery.

FINIS

Reflections *by Sir Charles Symonds*

Afterthoughts *by the Author*

Reflections

by Sir Charles Symonds, K.B.E., C.B., D.M. OXON., F.R.C.P.*

My own views on this subject have been influenced by the writings of Hughlings Jackson from whom I shall quote extracts relevant to the theme of your book.

Of the three alternatives Hughlings Jackson propounded for the relation of Consciousness to activities of the Highest Centers (1),† his own choice was the third—Concomitance (or psycho-physical parallelism). This I reject as being beyond my understanding. You conclude in favor of his first alternative, viz., "An immaterial agency is supposed to produce physical effects." This also I reject on the ground that it depends, scientifically speaking, upon negative conclusions. I am left with the second alternative "that activities of the highest centers and mental states are one and the same thing, or are different sides of the same thing."‡

My standpoint on this view agrees with that of Adrian (2) expressed in the 1966 symposium to which you refer. "The physiologist," he says, "is not forced to reject the old-fashioned picture of himself as an individual with a will of his own, for the position allows some validity to the introspective as well as the physiological account. It

* Formerly Neurologist at Guy's Hospital, London. These are his "Reflections" written after reading the prepublication manuscript.

† The bibliographic references may be found in a separate list of references at the end of this critique.

‡ See Note A in Afterthoughts.

admits that the two are incompatible, but does not maintain that they must always be so . . . Possibly our picture of brain events, or of human actions, may be changed so radically, that in the end they will account for the thinker as well as his thoughts." Your own conclusion and mine are alike in this—each requires an act of faith!

Hughlings Jackson has a good deal to say about the relation of consciousness to the highest nervous centers. He insists that every process of "mentation" must be associated with a corresponding activity in the nervous arrangements of the highest levels. "Consciousness,"* he says, "is not an unvarying independent entity." (3) Consciousness arises during activity of *some* of those of our highest nervous arrangements. His meaning in this context is further revealed by the statement, "Consciousness is a varying quantity—that is, we are from moment to moment differently conscious." (6) And again, "We spoke of the substrata of consciousness being the highest nervous arrangements. Yet to avoid misunderstanding, we pointed out explicitly . . . that we do not really sup-

* Hughlings Jackson makes it clear that for him consciousness has a wide meaning. "We have not got a consciousness *and* latest and highest mental states; we have only two names for one thing." (4) And again, "Whilst a man is having states of consciousness (in artificial analysis, Willing, Remembering, Reasoning and Feeling) there are correlatively with those purely psychical states, the physical things, slight discharges of nervous arrangements of his highest centers." (5)

For the clinician, loss of consciousness and impairment of mental activities are quite different. A patient who suffers from what we call dementia may by our standards be fully conscious. The man who is asleep is unconscious, but we do not judge him to be demented. Nevertheless in our discussions of the mind-brain relationship we are accustomed to take consciousness to represent the highest level of mentation.

pose there to be one fixed seat of consciousness. Now, if the expression be permitted, we shall speak of several highest nervous arrangements." (7)

His belief is that the pattern of these nervous arrangements is constantly changing. He says, "I do not mean that there are fixed nervous arrangements, answering to particular ideas, etc., in the anatomical substrata of subject consciousness;* there are only such nervous arrangements when the ideas etc. are actual. At other times the cells and fibres, in constant activity, keep up a state of general tension (nervous tonus); it is on the particular breaking of this strained equilibrium that nervous arrangements temporarily exist." (8)

Writing again of these nervous arrangements he says, "To speak of the anatomy, physiology, and pathology of loss of consciousness may seem strange; but we have already explained that the anatomical substrata of consciousness are only highly compound sensori-motor processes fundamentally like those of lower centres, differing in being the most special and complex of all. They have then, their anatomy, their physiology and their pathology." (9)

And in a footnote he adds, "Moreover the anatomo-physiological statement is not loss of consciousness, but, as previously explained, loss of correspondence of the organism as a whole with its environment. When we come to anatomy we shall point out that there is not one seat of consciousness, for consciousness is a varying quantity —that is we are from moment to moment differently conscious, we are continually changing correspondence

* In this passage Hughlings Jackson has been discussing the duality of consciousness—subjective and objective. Elsewhere he states his view that these are in reality two halves of the same thing.

with our environment." (10) Throughout his arguments there is insistence on the separation of consciousness, which is immaterial, from the nervous arrangements, which are physical; but occasionally he seems to depart for a moment from his own rule. For example in discussing anarthria, observed in a case of bulbar palsy, he notes that the patient can speak "internally," concluding that the cerebral motor nervous arrangements continue to represent articulatory movements, though the medullary centers and muscles have disappeared. He adds "these cerebral motor nervous arrangements have, as they have in healthy men, a psychical side." (11) On the hypothesis I have chosen, all nervous arrangements at Hughlings Jackson's highest level "have their psychical side." The anatomy, physiology, and pathology of these nervous arrangements are directly concerned with the mental phenomena dependent upon their activities. We are, therefore, justified in thinking, as far as we are able, about the anatomy, physiology, and pathology of the mind.*

Anatomy

Here I think we should inquire how low in the evolutionary scale can we detect, from behavior, what Hughlings Jackson called "mentation."

You mention the ant in this connection. I wonder if you had in mind the observation recorded by Thomas Belt in *The Naturalist in Nicaragua.* (12) I quote, "I shall conclude this long account of the leaf cutting ants with an instance of their reasoning powers. A nest was made near one of our tramways, and to get to the trees the ants had to cross the rails, over which the waggons

* See Note B in Afterthoughts.

were continually passing and repassing. Every time they came along a number of ants were crushed to death. They persevered in crossing for several days, but at last set to work and tunnelled under each rail. One day when the waggons were not running I stopped up the tunnels with stones; but although great numbers carrying leaves were thus cut off from the nest, they would not cross the rails, but set to work making fresh tunnels underneath them." As commentary upon this observation I quote W. H. Thorpe: "If we can see purposive behavior in animals or man, we have provisional grounds for believing that there is within the organism some sort of expectancy of the future, which entails or implies a capacity for ideation, an integration of ideas about past and future, and a temporal organization of ideas." (13) I know nothing about the nervous system of the ant, but it would seem that it contains a prototype of the anatomical substrata of mind.

J. Z. Young (14) has made a prolonged study of learning behavior in the octopus, and pursued in this creature the search for the neurones concerned, arriving at some interesting conclusions. I suggest that comparable studies should be made in other creatures before we jump to conclusions about the human mind. There is, I think, plenty of scope for this. Dolphins, birds, cats, dogs, and chimpanzees all manifest behavior showing that they have powers of reasoning. It may be objected that there is no evidence of self-awareness in these lower forms of life. But then there is no evidence that there is not. The cringing behavior of a dog detected by its master in some misdemeanor suggests that there may be.*

Your concept of a centrencephalic localization of Hughlings Jackson's highest level doubtless has more ana-

* See Note C in Afterthoughts.

tomical basis than I am aware of. I should have welcomed more details of this than appear in your monograph. For example, you state of sensory impulses that each stream comes to a first cellular interruption in the gray matter within the brain stem, but continues in a detour out to a second cellular interruption in the gray matter of the cerebral cortex. From there it returns directly to the target nucleus of cells within the gray matter of the higher brain-stem. I ask myself where is the anatomical verification for this last sentence. Incidentally I would point out that your statement takes no notice of the fact that the first cellular interruption is in the gray matter of the spinal cord, and that at this level there is complex interactivity of facilitation and inhibition, both local and centrifugal. (15, 16)

The anatomy of the reticular formation is another matter and, I suppose, more relevant to wakefulness; though this is really an aspect of consciousness.*

Physiology

The definition that Stanley Cobb gave me of consciousness was "a function of the brain in action." Function is a physiological term, and for this reason Hughlings Jackson would not allow its use for a description of mental activity; but on my hypothesis the definition is justified. It does not imply any localization of the function, which, I suggest, may be diffusely represented, varying both in space and time. As Hughlings Jackson said, we are differently conscious from one moment to another. It is a function, presumably, of synaptic activity, now here and now there. It seems to me more probable that its repre-

* See Note D in Afterthoughts.

sentation is in the cortex than in the diencephalon, having regard to the relative numbers of neurones available. That considerable areas of cortex may be injured, or removed, without loss of consciousness may depend upon a large amount of equipotentiality in what you call the uncommitted cortex.

The synaptic activity associated with consciousness is continuously present except during sleep. The explanation for this appears to be that the reticular formation in the brain stem in some way facilitates, or "drives" the higher centers, and that in sleep the activity of the reticular formation is inhibited. Here the relationship of consciousness to the brain-stem seems well established.

From an electroencephalogram it appears that neuronal activity in the cortex is constant even in sleep. The energy for all this neuronal activity is derived from the metabolism of glucose, for which the brain has a demand-for-size greater than that of any other organ. The crucial question, of course, is whether the *direction* of mental activity is itself a neural function. Such direction of mental activity appears to occur in the leaf-cutting ant, in which it is difficult to suppose that the direction is due to an immaterial agent.

Pathology

You say that injury or disease of the higher brain-stem always causes loss of consciousness. I think this may be true of acute lesions, e. g., trauma or vascular accident. But then the loss of consciousness resembles that of sleep.

Geoffrey Jefferson (17) in writing of concussion made this point, and coined the term "parasomnia" to describe the condition. The mind is in abeyance, but there are, he says, variations in the picture that can be regarded as "the arrest of a moment of awakening and the giving to it of duration." I wonder whether the state of your patient Landau, when you first saw him, may not have been one of parasomnia in the Jeffersonian sense?

Chronic disease of the higher brain-stem, as caused by tumors, does not, I think, as a rule, cause dementia, or loss of consciousness in the absence of increased intracranial pressure. When there *is* loss of consciousness it takes the form of parasomnia, or of akinetic mutism. Cairns (18) in his study of brain-stem tumors arrived at the same conclusion. However, dementia is constantly observed when the cerebral cortex is extensively affected by disease, as in the presenile dementias and the lipidoses. The traumatic dementias associated with widespread bilateral damage to white matter, without necessarily any brain-stem lesion (19), offer another example of localization other than centrencephalic.

One of the most significant events in the story of the mind-brain relationship was, I think, the occurrence of the obsessive-compulsive syndrome as a sequel of encephalitis lethargica. Because the observation is more than half a century old it is apt to be forgotten. I saw a great many of these patients, whose mental lives were preoccupied with compulsive thoughts, compulsive utterances, and compulsive behavior to an extent that often caused severe disability. Here were psychical events resulting from a physical cause, and one of the regions severely involved in this disease was the brain-stem, and especially the peri-aqueductal gray matter.

I imagine the symptoms I have described to have been caused by a defect of inhibitory control of neurones whose function is to drive the mental machinery—to use a crude analogy. This does not imply that the neurones comprising the anatomical substrata of mind are situated in the brain-stem. On the contrary, there is evidence to suggest that these neurones, those that were in a state of enhanced activity in consequence of the failure of inhibition, were far remote from the lesion.

The obsessive-compulsive syndrome, comparable with that of the postencephalitic patients, is quite commonly observed in persons who have not suffered from encephalitis, and is then sometimes subject to remission and relapse, suggesting a reversible biochemical lesion. When there is no remission bifrontal leucotomy may effect a cure. Of this there is no doubt. I have published a case (20) illustrating this sequence of events with a satisfactory follow-up over many years. I assume that in such cases it is the removal of the target-neurones, acted upon by the overactive centers in the brain stem, that is effective. It follows that these target neurones (being part of the anatomical substrata of mind) lie in the frontal cortex, not in the brain stem. I do not mean to imply that the frontal cortex is of special importance in this connection. I suppose the anatomical substrata of mind to be widely distributed throughout the cortex, and that ablation of other areas of cortex might have the same effect.

Then there is the problem of schizophrenia. Here we observe a disorder of mind with characteristic features—thought disorder, inappropriate affect, or lack of affect, forced thinking, delusions, hallucinations, etc., which surely has an organic basis, though as yet none has been discovered. The genetic factor in this disease is of outstanding importance. If one sibling is affected, the chances of another having the same disease is 14 percent as compared with a general incidence of 0.8 percent. Genes are physical things. The absence of any structural pathology suggests a biochemical disorder—perhaps an enzymatic defect. It may be argued that it is only the organ of mind that is diseased in schizophrenia, but introspective judgment, derived from the observation of many cases, leads me to the conviction that there is involvement of the thinker as well as his thoughts.

I used to discuss these problems with Russell Brain. Originally a dualist, he became converted to what he called a neutral monism, and this, I think, is a fair description of my own position.

I have said nothing about your fascinating observations on the interpretive-cortex, or on speech, which are of classical importance, as I do not think they are especially relevant to the matters I have discussed here.

References

1. JACKSON, J. H. 1931. *Selected Writings of John Hughlings Jackson*. Vol. II, p. 84. London: Hodder and Stoughton.
2. ADRIAN, E. D. 1966. In *Brain and Conscious Experience*, J. C. Eccles, ed. New York: Springer-Verlag, p. 238.
3. *Selected Writings*. Vol. I, p. 242.
4. *Selected Writings*. Vol. I, p. 289.
5. *Selected Writings*. Vol. II, p. 402.
6. *Selected Writings*. Vol. I, p. 205, footnote.
7. *Selected Writings*. Vol. I, p. 241.
8. *Selected Writings*. Vol. II, p. 98, footnote.
9. *Selected Writings*. Vol. I, p. 205.
10. *Selected Writings*. Vol. I, p. 205, footnote.
11. *Selected Writings*. Vol. II, p. 207.
12. BELT T. *The Naturalist in Nicaragua*. Everymans Library, p. 69.
13. THORPE, W. H. 1966. In *Brain and Conscious Experience*. J. C. Eccles, ed. New York: Springer-Verlag, p. 472.

14. Young, J. Z. 1965. *Proc. Roy. Soc. B.* 163:235.
15. Dawson, G. D. 1958. *Proc. Roy. Soc. Med.* 51:531.
16. Denny-Brown, D., E. J. Kirk, and N. Yanagisana, 1973. *J. Comp. Neur.* 151:175.
17. Jefferson, G. 1944. *B.M.J.* 1:1.
18. Cairns, H. 1952. 75:109.
19. Strich, S. J. 1961. *Lancet.* 2:443.
20. Symonds, Sir C. 1956. *Imprensa Medica.* Lisbon.

Afterthoughts by the Author

When the text of this monograph was completed I sent a copy to Sir Charles Symonds, a London neurologist for whom I have great respect and affection. He responded with a letter and a scholarly critique. I wrote him in reply as follows:

> Thank you for your letter and for the "Reflections on *The Mystery of the Mind*." When I mailed to you my manuscript by way of Adrian, I hoped you would do just this. But I hardly hoped for such a studied treatise. It is beautifully expressed. You say you have been deliberately critical. Yes, of course, that is what I wanted. It is only the truth that interests me, as it does you.
>
> Your observations supplement mine in a remarkable way. Some of them would inform the reader of my book and help him to understand what has been going on in the broad field of neurology. Your discussion of the philosophy of the great clinician, John Hughlings Jackson and your neurological survey of the relationship of "mental disease" due to brain abnormality will add perspective to this monograph—if you will agree to the proposal I am about to make to you:
>
> I would like to publish your critique just as you have sent it to me. I shall place it after the text. Your objections and questions would give me an opportunity to clarify what was not clear in my descriptions and to meet your objections if I can. To this end, I propose to follow your critique with a brief clarification, entitled "Afterthoughts."

I am hoping you will join me thus in this endeavor. Our friendship has been such a happy one!, beginning that summer when we were studying neuropathology and sat working with microscopes at the same bench in Godwin Greenfield's Laboratory at Queen Square. We have renewed it through the years, in your home and at Harvey Cushing's Clinic in Boston, and *chez nous* here in the Montreal Neurological Institute!

I shall make a few comments on your arguments, inserting a reference such as A, B, or C, etc. in your text to signify where each would apply. But I shall not attempt to answer you when I am persuaded that the text of the monograph has already done this.

Sir Charles agreed, and so my afterthoughts follow:

A.

You review Jackson's three alternative conclusions in regard to the brain-mind problem. Then you suppose that I am satisfied to adopt his first alternative, "an immaterial agency" producing "physical effects," for my own working hypothesis.

I am *not* satisfied to accept Jackson's hypothesis that you assign to me. First of all, I do not like the phrase "immaterial agency" since it seems to make assumptions yet to be examined. Secondly, I don't begin by a conclusion at all and I don't end by making a final and unalterable one. But I shall elaborate this point of view at the end of these afterthoughts.

As a matter of fact all through my experimental and exploratory career, I adopted the assumption that you (and Lord Adrian, you say) accept: "that activities of the highest centers and of mental states are one and the same thing."

That is the correct scientific approach for a neurophysiologist: to try to prove that the brain explains the mind and that mind is no more than a function of the brain. But during this time of analysis, I found no suggestion of action by a brain-mechanism that accounts for mind-action (Chapter 17). That is in spite of the fact that there is a highest brain-mechanism and that it seems to awaken the mind, as though it gave it energy, and seems itself to be used in turn by the mind as "messenger." Since I cannot explain the mind on the basis of your "assumption," I conclude that one must consider a second hypothesis: that man's being is to be explained by two fundamental elements.

B.

I am delighted to have your discussion of the thinking of Hughlings Jackson in regard to consciousness and what he called the "highest nervous arrangements." No one today could present the Gospel of Neurology according to John Hughlings Jackson so well as you. Don't think I am being flippant or disrespectful. I'm not. Certainly I am as much indebted to him as you. I think I am more so.

Jackson came to be London's great prophet of neurology because he read the meanings of epilepsy in terms of brain function. He surmised the truth, that in each epileptic seizure there is a discharge, a release of energy, in gray matter of the brain. He saw the result, in each case, as an illuminating experiment that could throw light on the functional "arrangements" within the brain.

It is a hundred years since the time of Jackson. We have been able to use stimulating electrodes that do the things that an epileptic discharge can do, i.e.: to activate or to arrest mechanisms of the brain, elaborating and adding to Jackson's surmises. With the help of conscious patients, we have charted the motor and sensory cortex, and added much detail. Areas of cortex, previously called silent, are now recognized as psychical. It is time to reconsider the data old and new. There is a physiology of epilepsy, too, as Jackson discovered. All this can now be amplified. It is time to apply this clearer understanding of epilepsy, that I have outlined, to the physiology of brain function. But I question whether, even now, we should speak of a "physiology of the mind," as Jackson does; not until we discover more about the nature of the mind.

C.

Surely you are not suggesting that investigators should solve the problems of "mentation" in organisms that are lower in the evolutionary scale than man, before approaching the problem in man! We have learned a great deal about the engram that is responsible for memory in man. It could not possibly have been discovered in lower forms that do not speak. For example, stimulation of the *interpretive-cortex* of a conscious man causes the record of the stream of consciousness to bring back to him the past, and he speaks during the stimulation to tell us of it.[19,20,21,22] (See *Proceedings of the Royal Society of Medicine*, August, 1968, "Engrams in the Human Brain.") *Nothing* is more relevant to an understanding of consciousness than what we have learned from stimulation of the interpretive-cortex.

The beautiful story of Belt's leaf-cutting ants that you have retold seems, to me, to show clearly that there is self-awareness in the ant. That ant who carries his leaf to the rail and stops, finding the hole closed underneath the rail, must be aware of himself and his predicament if he begins to make a new hole.

Life made its appearance on this planet a long while ago and a long while after that, organisms, whose behavior suggests self-awareness or consciousness, made their first appearance. One can only suppose that it was then that mind made its appearance on earth whether it was as a function or as a related element. When I speak thus of mind and brain I am assuming that a mind existed only in relation to a brain.

After the creation of matter and the appearance of life, consciousness made its appearance on earth quite late in the universal calendar. We do well to reserve opinion of the past until we understand man, who can be so much more easily studied in the present. Explorers in each field of neuroscience should push ahead wherever valid clues present themselves. But exploration of the mind and brain of man throws light on all other creatures.

Such work as this monograph is necessarily hypothetical and prophetic. It is a study of neurophysiology. Some day anatomists, chemists, and physicists will explain the "how" of many things, no doubt. Meanwhile the mechanisms of brain-action that make sensation, movement, language, perception, learning, memory, and consciousness possible can be understood by us with ever-increasing certainty, even before we comprehend the "how" of a neuron's conduction of electrical potentials, and before we complete the mapping of circuits within the brain of lower forms.

I spent six months working on the microscopic anatomy of the mammalian brain in Madrid with del Río-Hortega and the great Ramón y Cajal, whom Sherrington considered the greatest neuroanatomist of all time. One of my Spanish associates in the Madrid laboratories had been started by Cajal on a study of the ant's brain years before. The disciple complained to me that Cajal had hoped originally to discover a simpler brain in the ant. Not so. It seemed far more complicated than expected. Consequently Cajal turned back to the mammalian brain, the brain that you and I are using, as best we can, in this effort to understand man. You and I can only be thankful that Cajal did turn back, even though he left his assistant unaided in his effort to understand the meaning, in terms of function, of the multiform circuits encountered in the ant's brain.

D.

Yes, I agree that a first cellular interruption for somatic sensation does occur in the spinal cord, and I am sure it is important for spinal reflexes. But we are considering a higher integration. I was thinking of the arrival of each afferent stream of impulses at a higher level as having a first cellular interruption in the diencephalon. With the exception of pain, the other important sensory streams (visual, somatic, and auditory) make a detour to special areas of gray matter on the convolutions of the cortex where they come to a second cellular interruption.

You ask for "anatomical verification" of my statement that from there, each sensory stream "returns directly to the target-nucleus of cells within the gray matter of the higher brain-stem." This question is one that F. M. R. Walshe asked. To some extent the answer is self-evident, since each convolution is an outbudding of some thalamic nucleus in the diencephalon, as shown by Earl Walker's early anatomical studies.

The "target-nucleus," as I have called it, must be within the integrative complex. No one, as yet, can describe the detailed microscopic *anatomy* of integration with its automatic inhibitions and activations. No one can map the intricacies of man's marvelous computer. But automatic integration *does* go on. Man *has* a computer. There *is* reflex action and there is reasoned action. There *is* a highest brain-mechanism and it *can be inactivated selectively* by epileptic discharge within the diencephalon.

I can also answer your question by giving you the conclusions and the evidence of physiology, epilepsy, and neurosurgery. This, too, is neurology.

During surgical treatment for epilepsy, conscious patients have allowed neurosurgeons to establish exact frontiers for all these sensory areas on the human cortex. This has been verified, as you well know, by using a stimulating electrode and by excising convolutions here or there in the hope of cure.

As a neurosurgeon who has removed convolutions down to where the stem or white matter of the convolution is attached to the diencephalon in hundreds of cases, the answer to your question is quite easy, so easy in fact that I am startled to realize that I did not present it clearly.*

* Take the postcentral gyrus: Through it flows sensory data that make discriminatory sensation possible. If the gyrus behind it is removed or if the precentral gyrus, which is in front of it, is removed completely, or if both of them are removed, sensation in the fingers of the opposite hand is not lost. Point localization is still possible, and the sense of position of a finger in space is intact. Therefore, the information has reached the target, which must be in the diencephalon. If, however, the postcentral gyrus is selectively removed and the white-matter tracts that connect the cortex with a thalamic nucleus of the diencephalon are cut, discriminative sensation is lost in the fingers of the opposite side. The same can be said of the visual cortex. As long as its connection to the diencephalon is preserved, vision in the opposite visual field is preserved.

Obviously the somatic-sensory data and the visual data required for conscious voluntary motor activity do enter the diencephalon.

Next there is clear evidence that the *outgoing stream of neurone potentials that initiates voluntary activity* comes from the diencephalon and enters the precentral gyrus on its way to the muscles of the body. This initiating stream of impulses cannot come to the precentral gyrus from other parts of the brain along so-called "association tracts," because wide removal of other parts of cortex down to diencephalon does not interfere with this voluntary control.*

I have referred to the system of connections that must serve the integrative activity that comes between the sensory input, into the diencephalon, and the motor output as the *centrencephalic* system. Within that integrating system there is an automatic computer and a mechanism that makes consciousness possible. When the specialized areas of the cerebral cortex are in action, some of them at least are as much involved in the neurone action related to consciousness as is the diencephalon.

General Conclusion

To suppose that consciousness or the mind has localization is a failure to understand neurophysiology. The great mathematician and philosopher, René Descartes (1596–1650), made a mistake when he placed it in the pineal gland. The amusing aspect is that he came so close to that part of the brain in which the essential circuits of the highest brain-mechanism must be active to make consciousness possible.

* More complete substantiation of this and other evidence is to be found in "Mechanisms of Voluntary Movement."[24]

Centrencephalic integrative action serves the purposes of two semi-separable mechanisms, (a) the highest brain-mechanism, and (b) the automatic sensory-motor brain-mechanism (or brain-computer). Perhaps I should say that these two mechanisms are the two parts of this final centrencephalic integration. The centrencephalic system has a major direct connection with the new portions of the frontal lobes and of the temporal lobes that form such a large proportional addition in passing up the mammalian scale from ape to man. The anatomical verification for this statement is to be found in the recent anatomical studies of Walle Nauta.[12] The connection is between anterior frontal cortex and hypothalamus on the one hand. It is also between antero-inferior temporal cortex and the medio-dorsal nucleus of the thalamus on the other. The evidence from our studies of epileptic patterns suggests that gray matter in the highest brain-mechanism had direct connection with these parts of frontal and temporal lobes while the gray matter of the *automatic* sensory-motor mechanism is directly connected to the various areas of sensory and motor cortex. One may assume that the highest brain-mechanism makes its major connections with the sensory and motor cortex only through the automatic mechanism of the diencephalon.*

Let me reconsider: Sensory targets must be in close relation (i.e., closely connected to) to the mechanism from which voluntary motor control emerges to the periphery. That means that it is in the higher brain-stem. It is a clinical and physiological hypothesis that has now been verified by long experience with brain stimulation

* This anatomical substantiation[12] of the basis for the action of the centrencephalic system was announced by Nauta (to my delight) in his Hughlings Jackson Lecture at the Montreal Neurological Institute, May 8, 1974.

and convolution removal. But let me look back for a moment: For years it had been assumed that sensory data were taken to the cerebral cortex, and that conscious behavior was controlled by the cortex, and that the "association" nerve fibers that run over the surface of the cortex somehow accounted for all that takes place between sensory input and motor output. Jackson suggested the highest level of integration might be in the anterior frontal lobes for a time. But in the face of clinical experience, he let the suggestion lapse. The importance of his teaching was that he knew it would be discovered somewhere and he had ideas about its physiological organization that had come to him from the study of epilepsy.

My curiosity about the mind-brain relationship began during the 1920s in a major study of decerebrate cats. I joined Cuthbert Bazett in this study while in Sherrington's laboratory. We removed the higher brain-stem and hemisphere of these animals, preserving the lower brain-stem and spinal cord. After operation, they had no consciousness, no mind. They were completely automatic reflex "preparations." As pointed out in the Preface, life was there and reflex mechanisms were there, but there was no evidence of mind. After turning from physiology to neurology and neurosurgery, I eventually had the opportunity of working as a neurosurgeon with an excellent team and facilities for study in the Montreal Neurological Institute. It opened its doors in 1934. In 1936 I was called upon to give the Harvey Lecture at the Academy of Medicine in New York.[14] I chose "The Cerebral Cortex and Consciousness" for my title. I presented the obvious conclusions that had been forced upon me in the first years of neurosurgical practice, notably the evidence derived from therapeutic brain removal, exploratory electrical stimulation, and the obvious relationship of unconsciousness to injury to the diencephalon, which was evident in clinical practice.

As I pointed out in Chapter 5 (page 18), my Harvey Lecture was important in the evolution of my own clinical understanding. It was so obvious that the integration within the central nervous system, which makes consciousness possible, did not take place in the cerebral cortex. It was not to be found in the new brain at all, but in the old brain. It was "below the cortex and above the midbrain." Much that I surmised in 1936 was proven over and over again in clinical cases, during the years that followed.

I closed the Harvey Lecture with these words:

> This discussion has . . . been concerned with the localization of the "place of understanding," and by "place" is meant the location of those neuronal circuits which are most intimately associated with the initiation of voluntary activity and with the sensory summation prerequisite to it.

In 1952 I published a paper, "Epileptic Automatism and the Centrencephalic Integrating System."[15] It was clear that epileptic discharge passed directly from anterior frontal cortex and also from anterior temporal cortex to the diencephalon, where the interfering discharge inactivated a highest brain-mechanism that was most closely related to consciousness. Thinking of a new word to suggest that final integrative action took place centrally in the diencephalon where primary epileptic inactivation as well as injury produced loss of consciousness, I proposed the phrase *centrencephalic integration*. I could only presume the existence of the anatomical connections that Nauta has now demonstrated. I assumed, and still assume, that each sensory area (on whatever convolution it is located in the cerebral cortex) sends its stream of data onward into that part of the diencephalon from which the cortical convolution in question is the structural outgrowth. Now that examination of the neurophysiology of epilepsy makes it possible to recognize (a) the *automatic sensory-motor mechanism* in the diencephalon, and (b) the *highest brain-mechanism*, there is good evidence that the afferent stream of sensory information leads directly into (a) the automatic sensory-motor mechanism. From there, unless it is blocked by inhibition, it passes on to the level of conscious awareness. That is to say, it passes on to (b) the highest brain-mechanism.

Without the anatomical detail, I give you indirect evidence, which is clinical and physiological: Voluntary behavior is controlled by a stream of neuronal potentials that emerges from the diencephalon. Complicated maneuvers have a cellular interruption in the precentral motor gyrus, while gross control is carried out by direct nerve connections to the spinal cord. (A "stroke" that blocks conduction through the precentral gyrus makes it impossible to play the piano, but leaves the capacity to make gross pawlike movements.) From the precentral gyrus the stream of potentials goes out along the well known pathway to the muscles of the body and results in complicated voluntary action. The data, from sensory inflow and from mind-action, *must* reach the neuronal circuits of the diencephalon. (See Chapters 3, 4, 5, and 6.)

And so I come to my final reconsideration: I worked as a scientist trying to prove that the brain accounted for the mind and demonstrating as many brain-mechanisms as possible hoping to show *how* the brain did so. In presenting this monograph I do not begin with a conclusion and I do not end by making a final and unalterable one.

Instead, I reconsider the present-day neurophysiological evidence on the basis of two hypotheses: (a) that man's being consists of one fundamental element, and (b) that it consists of two. I take the position that the brain-mechanisms, which we (my many colleagues and I all around the world), are working out, would, of course, have to be employed on the basis of either alternative. In the end I conclude that there is no good evidence, in spite of new methods, such as the employment of stimulating electrodes, the study of conscious patients and the analysis of epileptic attacks, that the brain alone can carry out the work that the mind does. I conclude that it is easier to rationalize man's being on the basis of two elements than on the basis of one. But I believe that one should not pretend to draw a final scientific conclusion, in man's study of man, until the nature of the energy responsible for mind-action is discovered as, in my own opinion, it will be.

Thus, let me state again that, working as a scientist all through my life, I have proceeded on the one-element hypothesis. That is really the same as the Jacksonian alternative that Symonds and Adrian seem to have chosen, i.e., "that activities of the highest centers and mental states are one and the same thing, or are different sides of the same thing."

I am glad to see that Professor Hendel has referred to dualism in his Foreword. In any case, as a scientist, I reject the concept that one must be either a monist or a dualist because that suggests a "closed mind." No scientist should begin thus, nor carry on his work with fixed preconceptions. And yet, since a final conclusion in the field of our discussion is not likely to come before the youngest reader of this book dies, it behooves each one of us to adopt for himself a personal assumption (belief, religion), and a way of life without waiting for a final word from science on the nature of man's mind.

In ordinary conversation, the "mind" and "the spirit of man" are taken to be the same. I was brought up in a Christian family and I have always believed, since I first considered the matter, that there was work for me to do in the world, and that there is a grand design in which all conscious individuals play a role. Whether there is such a thing as communication between man and God and whether energy can come to the mind of a man from an outside source after his death is for each individual to decide for himself. Science has no such answers.

Bibliography

1. Adrian, E. D. 1966. Consciousness, in *Brain and Conscious Experience*, J. C. Eccles, ed. New York: Springer-Verlag.
2. Feindel, W. and W. Penfield. 1954. Localization of discharge in temporal lobe automatism. *Arch. Neurol. and Psychiat.* 72:605–630.
3. Jackson, J. H. 1873. On the anatomical, physiological and pathological investigation of the epilepsies. *West Riding Lunatic Asylum Medical Reports.* 3:315–339.
4. ——— 1931. *Selected Writings of John Hughlings Jackson.* Vol. 1, On Epilepsy and Epileptiform Convulsions, J. Taylor, ed. London: Hodder and Stoughton.
5. James, W. 1910. *The Principles of Psychology.* New York: Holt.
6. Hippocrates. W. Jones and E. Withington, eds. The Loeb Classical Library, 4 volumes. Vol. 2, *The Sacred Disease,* pp. 127–185. Cambridge: Harvard Univ. Press. Also London: Heinemann, 1952–1958.
7. Lashley, K. S. 1960. *The Neuropsychology of Lashley; Selected Papers of.* F. A. Beach et al., eds. New York: McGraw-Hill.
8. Magoun, H. 1952. The ascending reticular activating system. *A. Res. Nervous and Mental Disease, Proceedings* (1950) 30:480–492.
9. Moruzzi, G. and H. Magoun. 1949. Brain stem reticular formation and activation of the EEG. *Electroenceph. Clin. Neurophysiol.* 1:455–473.
10. Mullan, S. and W. Penfield. 1959. Illusions of comparative interpretation and emotion. *Arch. Neurol. and Psychiat.* 81:269–284.

11. PAVLOV, I. P. 1927. *Conditioned Reflexes: An Investigation of the Physiological Activity of the Cerebral Cortex*. G. Anrep, trans. and ed. London: Oxford Univ. Press.
12. NAUTA, W. 1971. The problem of the frontal lobe: a reinterpretation. *J. Psychiat. Res.* 8:167–187.
13. PENFIELD, W. 1930. Diencephalic autonomic epilepsy. *Arch. Neurol. and Psychiat.* 22:358–374.
14. ——— 1938. The cerebral cortext in man. The cerebral cortex and consciousness. *Arch. Neurol. and Psychiat.* 40:417–422. Also in French, Prof. H. Piéron, trans. in *L'Année Psychologique*. 1938. Vol. 39.
15. ——— 1952. Epileptic automatism and the centrencephalic integrating system. *A. Res. Nervous and Mental Disorders, Proceedings* (1950) 30:513–528.
16. ——— 1952. Memory mechanisms. *Arch. Neurol. and Psychiat.* 67:178–191.
17. ——— 1954. The permanent record of the stream of consciousness. Proc. XIV Int. Congr. Psychol., Montreal. *Acta Psychologica*. 11:47–69. 1955.
18. ——— 1958. *The Excitable Cortex in Conscious Man*. The 5th Sherrington Lecture. Liverpool: Liverpool Univ. Press. Also Springfield, Ill.: Charles C Thomas.
19. ——— 1959. The interpretive cortex. *Science*. 129:-1719–1725.
20. ——— 1968. Engrams in the human brain. *Proc. Roy. Soc. Med.* 61:831–840. (Gold Medal Lecture.)
21. ——— 1969. Consciousness, memory and man's conditioned reflexes. In *On the Biology of Learning*. K. Pribriam, ed. New York: Harcourt, pp. 129–168.
22. ——— 1969. Epilepsy, neurophysiology, and some brain mechanisms related to consciousness. In *Basic Mechanisms of the Epilepsies*, H. H. Jasper et al., eds. Boston: Little, Brown.

23. PENFIELD, W. and J. EVANS. 1935. The frontal lobe in man: A clinical study of maximum removals. *Brain.* 58:115–138.
24. ———. 1954. Mechanisms of voluntary movement. *Brain.* 77:1–17.
25. ——— and H. JASPER. 1954. *Epilepsy and the Functional Anatomy of the Human Brain.* Boston: Little, Brown.
26. ——— and K. KRISTIANSEN. 1951. *Epileptic Seizure Patterns.* Springfield, Ill.: Charles C Thomas.
27. ——— and G. MATHIESON. 1974. Memory. An autopsy and a discussion of the role of the hippocampus in experiential recall. *J.A.M.A. Archives of Neurology.* 31:145–154.
28. ——— and P. PEROT. 1963. The brain's record of auditory and visual experience. A final summary and discussion. *Brain.* 86:595–696.
29. ——— and T. RASMUSSEN. 1950. *The Cerebral Cortex of Man.* New York: Macmillan.
30. ——— and L. ROBERTS. 1959. *Speech and Brain-Mechanisms.* Princeton: Princeton Univ. Press. Also New York: Atheneum, 1966.
31. SHERRINGTON, C. S. 1940. *Man—On His Nature.* The Gifford Lectures, 1937–1938. Cambridge: Cambridge Univ. Press.
32. ——— 1947. Foreword to a new edition of *The Integrative Action of the Nervous System.* (Originally published in 1906.) Cambridge: Cambridge Univ. Press.

Index

Library of Congress Cataloging in Publication Data

Penfield, Wilder, 1891-1976
The mystery of the mind.

Bibliography: p. 116
1. Brain. 2. Consciousness. I. Title.
QP376.P39 612'.82 74-25626
ISBN 0-691-08159-X (hardcover edn.)
ISBN 0-691-02360-3 (paperback edn.)